PROF. MSDOSS MATHS BOOK SERIES IV

CALCULUS I
DIFFERENTIAL CALCULUS
Formulae Practice Workbook - Vol II

By Prof. M. Subbiah Doss ©

Author's email-id : subdoss2014@gmail.com

First Edition : April, 2017

Pages : 64

THAASU PUBLICATIONS

14, II CROSS STREET

VISWANATHAPURAM

MADURAI-625014. INDIA

Preface

After successfully completing the practice given in volume I, now we are ready to solve problems in differential calculus. Different methods are discussed in this workbook. Problems are categorized in ten different units. A self-evaluation test is provided for practice.

The units 10.1 to 10.4 usually appear at the beginning of any text book. But, for our convenience, here we have discussed these units at the end.

Explanation to the following doubts raised by the students is given after self evaluation test. They are:

1. Whether the symbol $\frac{dy}{dx}$ is a 'single thing' or it is $y \div dx$ (a ratio)?

2. $\Delta x \to 0$ means $\Delta x = 0$ or not?

3. We have proved 'If $y = \sin x$ then $\frac{dy}{dx} = \cos x$' and other similar results in volume II using the 'first principle technique'. But these results seem to have no life at all!

Try to recollect the Differential calculus formulae given at the beginning of this workbook. You can do this easily.

The practice gained by you in **'Differential Calculus Formulae Practice Workbook'- Prof.MSDOSS book series I** helps you to remember and recollect all these formulae without any doubt.

With the confidence gained, you can easily solve the problems by just following the methods available here. **Wish you all the best!**

Note: It is advisable to work with Differential Calculus volume II after completing Differential calculus volume I. You will find some lengthy and challenging problems in this workbook. However, with the practice gained in Differential Calculus volume I, one could solve these problems with ease. ☺

Contents

Revision of the formulae discussed in volume I	1
Unit 1 - Derivatives of functions involving sum and / or difference	3
Unit 2 - More problems in 'function of a function' (chain) rule	6
Unit 3 - Product rule	10
Unit 4 - Quotient rule	16
Unit 5 - Differentiation of Implicit functions	21
Unit 6 - Parametric form	26
Unit 7 - Substitution method	30
Unit 8 - Logarithmic differentiation	35
Unit 9 - Derivative of a function w.r.to another function	44
Unit 10 - 1. Differentiation Technique	47
2. Derivative of Inverse trigonometric functions	
3. Derivative of hyperbolic functions	
4. Derivative of inverse hyperbolic functions	
Self evaluation test	56
To the students ……..	62

CALCULUS I – Differential Calculus Formulae Practice Workbook Vol II

Recollect the following formulae

If $y = x^n$ then $\dfrac{dy}{dx} = nx^{n-1}$	If $y = x$ then $\dfrac{dy}{dx} = 1$
If $y = \dfrac{1}{x}$ then $\dfrac{dy}{dx} = -\dfrac{1}{x^2}$	If $y = \sqrt{x}$, then $\dfrac{dy}{dx} = \dfrac{1}{2\sqrt{x}}$

If $y = e^x$ then $\dfrac{dy}{dx} = e^x$ If $y = a^x$ then $\dfrac{dy}{dx} = a^x \log_e a$
If $y = \log_e x$ then $\dfrac{dy}{dx} = \dfrac{1}{x} \log_e e = \dfrac{1}{x}$
If $y = \log_a x$ then $\dfrac{dy}{dx} = \dfrac{1}{x} \log_a e$
If $y = \sin x$ then $\dfrac{dy}{dx} = \cos x$
If $y = \cos x$ then $\dfrac{dy}{dx} = -\sin x$
If $y = \tan x$ then $\dfrac{dy}{dx} = \sec^2 x$
If $y = \cot x$ then $\dfrac{dy}{dx} = -\csc^2 x$
If $y = \sec x$ then $\dfrac{dy}{dx} = \sec x \tan x$
If $y = \csc x$ then $\dfrac{dy}{dx} = -\csc x \cot x$

If $y = k$ (constant) then $\dfrac{dy}{dx} = 0$
If $y = ku$ (k constant, u is a function of x) then $\dfrac{dy}{dx} = k \dfrac{du}{dx}$
If $y = u + v$ (u, v are functions of x) then $\dfrac{dy}{dx} = \dfrac{du}{dx} + \dfrac{dv}{dx}$
If $y = u - v$ (u, v are functions of x) then $\dfrac{dy}{dx} = \dfrac{du}{dx} - \dfrac{dv}{dx}$
If $y = u.v$ (u, v are functions of x) then $\dfrac{dy}{dx} = u.\dfrac{dv}{dx} + v.\dfrac{du}{dx}$
If $y = u.v.w$ (u, v, w are functions of x) then $\dfrac{dy}{dx} = uv\dfrac{dw}{dx} + uw\dfrac{dv}{dx} + vw\dfrac{du}{dx}$
If $y = \dfrac{u}{v}$ (u, v are functions of x) then $\dfrac{dy}{dx} = \dfrac{v.\dfrac{du}{dx} - u.\dfrac{dv}{dx}}{v^2}$

If $y = \sin^{-1}(x)$ then $\dfrac{dy}{dx} = \dfrac{1}{\sqrt{1-x^2}}$

If $y = \cos^{-1}(x)$ then $\dfrac{dy}{dx} = -\dfrac{1}{\sqrt{1-x^2}}$

If $y = \tan^{-1}(x)$ then $\dfrac{dy}{dx} = \dfrac{1}{1+x^2}$

If $y = \cot^{-1}(x)$ then $\dfrac{dy}{dx} = -\dfrac{1}{1+x^2}$

If $y = \sec^{-1}(x)$ then $\dfrac{dy}{dx} = \dfrac{1}{x\sqrt{(x)^2-1}}$

If $y = \operatorname{cosec}^{-1}(x)$ then $\dfrac{dy}{dx} = -\dfrac{1}{x\sqrt{(x)^2-1}}$

If $y = \sinh(x)$ then $\dfrac{dy}{dx} = \cosh(x)$

If $y = \cosh(x)$ then $\dfrac{dy}{dx} = \sinh(x)$

If $y = \tanh(x)$ then $\dfrac{dy}{dx} = \operatorname{sech}^2(x)$

If $y = \coth(x)$ then $\dfrac{dy}{dx} = -\operatorname{cosech}^2(x)$

If $y = \operatorname{sech}(x)$ then $\dfrac{dy}{dx} = -\operatorname{sech}(x)\tanh(x)$

If $y = \operatorname{cosech}(x)$ then $\dfrac{dy}{dx} = -\operatorname{cosech}(x)\coth(x)$

If $y = \sinh^{-1}(x)$ then $\dfrac{dy}{dx} = \dfrac{1}{\sqrt{1+x^2}}$

If $y = \cosh^{-1}(x)$ then $\dfrac{dy}{dx} = \dfrac{1}{\sqrt{x^2-1}}$, $x > 1$

If $y = \tanh^{-1}(x)$ then $\dfrac{dy}{dx} = \dfrac{1}{1-x^2}$, $|x| < 1$

If $y = \coth^{-1}(x)$ then $\dfrac{dy}{dx} = \dfrac{1}{1-x^2}$, $|x| > 1$

If $y = \operatorname{sech}^{-1}(x)$ then $\dfrac{dy}{dx} = \dfrac{-1}{x\sqrt{1-x^2}}$

If $y = \operatorname{cosech}^{-1}(x)$ then $\dfrac{dy}{dx} = -\dfrac{1}{|x|\sqrt{x^2+1}}$

CALCULUS I – Differential Calculus Formulae Practice Workbook Vol II

Unit 1 Derivatives of functions involving sum and / or difference

Examples

1. Differentiate $4x^3 + 3x^2 - 14x - 8$ w.r.to x

   Let $y = 4x^3 + 3x^2 - 14x - 8$ then $\dfrac{dy}{dx} = 12x^2 + 6x - 14$

2. Differentiate $5e^x + 3\sin x - 2\tan x - \dfrac{1}{x}$ w.r.to x

   Let $y = 5e^x + 3\sin x - 2\tan x - \dfrac{1}{x}$ then

   $\dfrac{dy}{dx} = 5e^x + 3\cos x - 2\sec^2 x + \dfrac{1}{x^2}$

3. If $y = (x+3)(x^2 - 1)$, find $\dfrac{dy}{dx}$

   $y = (x+3)(x^2 - 1) = x^3 + 3x^2 - x - 3$

   Differentiating w.r.to x, $\dfrac{dy}{dx} = 3x^2 + 6x - 1$

4. Find $\dfrac{d}{dx}\left(\dfrac{x+1}{\sqrt{x}}\right)$; $\dfrac{x+1}{\sqrt{x}} = \dfrac{x}{\sqrt{x}} + \dfrac{1}{\sqrt{x}} = \sqrt{x} + x^{-\frac{1}{2}}$

   $\dfrac{d}{dx}\left(\dfrac{x+1}{\sqrt{x}}\right) = \dfrac{d}{dx}(\sqrt{x} + x^{-\frac{1}{2}}) = \dfrac{d}{dx}(\sqrt{x}) + \dfrac{d}{dx}(x^{-\frac{1}{2}}) = \dfrac{1}{2\sqrt{x}} - \dfrac{1}{2x^{3/2}}$

5. If $f(x) = \dfrac{(x+1)(x^2-3)}{x}$, find $f'(1)$

   $f(x) = \dfrac{x^3 + x^2 - 3x - 3}{x} = x^2 + x - 3 - \dfrac{3}{x}$

   Differentiating w.r.to x, we get $f'(x) = 2x + 1 + \dfrac{3}{x^2}$; $f'(1) = 6$.

Exercise :

1. Differentiate $\sec x + 2\cos x - \sqrt{x}$ w.r.to x

   Let $y = \sec x + 2\cos x - \sqrt{x}$ then

   $\dfrac{dy}{dx} = \sec x . \tan x - 2\sin x - \dfrac{1}{2\sqrt{x}}$

2. Differentiate $\sqrt{1+\sin 2x}$ w.r.to x

   Let $y = \sqrt{1+\sin 2x} = \sqrt{\sin^2 x + \cos^2 x + 2\sin x \cdot \cos x} = \sin x + \cos x$

   $\dfrac{dy}{dx} = \cos x - \sin x$

3. If $f(x) = x(x+1)(x-3)$, find $f'(x)$

   $f(x) = x(x+1)(x-3) = x^3 - 2x^2 - 3x$

   Differentiating w.r.to x, we get $f'(x) = 3x^2 - 4x - 3$

4. If $y = \sqrt[3]{x} - 2\log x + \sin^{-1}(\sin 3x)$, find y'

   $y = \sqrt[3]{x} - 2\log x + \sin^{-1}(\sin 3x) = x^{\frac{1}{3}} - 2\log x + 3x$

   Differentiating w.r.to x, $y' = \dfrac{1}{3}x^{-\frac{2}{3}} - \dfrac{2}{x} + 3$

5. Find $\dfrac{d}{dx}(3\tan^{-1}x - 2\sec^{-1}x + 5)$

   $\dfrac{d}{dx}(3\tan^{-1}x - 2\sec^{-1}x + 5) = 3\dfrac{d}{dx}(\tan^{-1}x) - 2\dfrac{d}{dx}(\sec^{-1}x) + \dfrac{d}{dx}(5)$

   $= \dfrac{3}{1+x^2} - \dfrac{2}{x\sqrt{x^2-1}}$

Find the errors if any :

1. If $y = \sqrt{x}(x^2+2)$, find $\dfrac{dy}{dx}$

   $y = \sqrt{x}(x^2+2) = x^{\frac{5}{2}} + 2\sqrt{x}$

   Differentiating w.r.to x, we get $\dfrac{dy}{dx} = \dfrac{5}{2}x^{\frac{3}{2}} + \dfrac{1}{2\sqrt{x}}$

2. If $f(x) = 4\sin x - \sin 2x$, find $f'(\dfrac{\pi}{2})$

   $f(x) = 4\sin x - \sin 2x$; Differentiating w.r.to x,

   $f'(x) = 4\cos x - 2\cos 2x$ and $f'(\dfrac{\pi}{2}) = 4\cos\dfrac{\pi}{2} - 2\cos\pi = 2$

3. Differentiate $\text{cosech } x - 3\tanh^{-1}x - 9$ w.r.to x

   Let $y = \text{cosech } x - 3\tanh^{-1}x - 9$ then

CALCULUS I – Differential Calculus Formulae Practice Workbook Vol II

$$\frac{dy}{dx} = -\operatorname{cosech} x \cdot \coth x - \frac{3}{1-x^2}$$

4. If $y = (x + \frac{1}{x})^2$ find $\frac{dy}{dx}$

$y = x^2 + \frac{1}{x^2} + 2$; Differentiating w.r.to x, $\frac{dy}{dx} = 2x - \frac{2}{x^3}$

5. If $y = \frac{2x+3}{\sqrt{x}}$ find $\frac{dy}{dx}$

$y = \frac{2x+3}{\sqrt{x}} = 2\sqrt{x} + 3x^{-\frac{1}{2}}$; Differentiating w.r.to x, $\frac{dy}{dx} = \frac{1}{\sqrt{x}} - \frac{3}{2} x^{-\frac{3}{2}}$

Corrected solution

1. $\frac{dy}{dx} = \frac{5}{2} x^{\frac{3}{2}} + \frac{1}{\sqrt{x}}$ 2,3,4,5 – No error

Do it yourself :

1. Differentiate $3 \log_5 x - \operatorname{cosec} x - 4 \cot^{-1} x$ w.r.to x

2. If $f(x) = \frac{a+x}{ax}$, find $f'(x)$

3. If $y = \frac{(x-3)(2x^2-4)}{x}$, find $\frac{dy}{dx}$

4. Find $\frac{d}{dx}[(x - \frac{1}{x})^3]$

5. ♣

♣ Why not you frame a problem and solve it? You can do it!

Answers :

1. $\frac{3}{x} \log_5 e + \operatorname{cosec} x \cdot \cot x + \frac{4}{1+x^2}$ 2. $-\frac{1}{x^2}$ 3. $4x - 6 + \frac{12}{x^2}$

4. $3(x^2 - 1 - \frac{1}{x^2} + \frac{1}{x^4})$ 5. ……………(write your answer here)

Unit 2 : More problems in 'function of a function'(chain) rule

Examples

1.1 If $y = \tan \sqrt{x}$,

then $\dfrac{dy}{dx} = \sec^2\sqrt{x} \cdot \dfrac{1}{2\sqrt{x}} = \dfrac{\sec^2\sqrt{x}}{2\sqrt{x}}$

1.2 If $y = \log(\tan \sqrt{x})$,

then $\dfrac{dy}{dx} = \dfrac{1}{\tan \sqrt{x}} \cdot \sec^2\sqrt{x} \cdot \dfrac{1}{2\sqrt{x}} = \dfrac{\sec^2\sqrt{x}}{2\sqrt{x} \cdot \tan \sqrt{x}}$

1.3 If $y = \sin^{-1}(\log(\tan \sqrt{x}))$,

then $\dfrac{dy}{dx} = \dfrac{1}{\sqrt{1-(\log(\tan \sqrt{x}))^2}} \cdot \dfrac{1}{\tan \sqrt{x}} \cdot \sec^2\sqrt{x} \cdot \dfrac{1}{2\sqrt{x}}$

$= \dfrac{\sec^2\sqrt{x}}{2\sqrt{x} \cdot \tan \sqrt{x} \cdot \sqrt{1-(\log(\tan \sqrt{x}))^2}}$

2.1 If $y = \sin 5x$,

then $\dfrac{dy}{dx} = \cos 5x \cdot 5 = 5 \cos 5x$

2.2 If $y = \tan^{-1}(\sin 5x)$,

then $\dfrac{dy}{dx} = \dfrac{1}{1+(\sin 5x)^2} \cdot \cos 5x \cdot 5 = \dfrac{5 \cos 5x}{1+(\sin 5x)^2}$

2.3 If $y = \tanh(\tan^{-1}(\sin 5x))$,

then $\dfrac{dy}{dx} = \text{sech}^2(\tan^{-1}(\sin 5x)) \cdot \dfrac{1}{1+(\sin 5x)^2} \cdot \cos 5x \cdot 5$

$= \dfrac{5 \cos 5x \cdot \text{sech}^2(\tan^{-1}(\sin 5x)}{1+(\tan 5x)^2}$

3.1 If $y = \log(x^2 + 1)$, then $\dfrac{dy}{dx} = \dfrac{2x}{x^2+1}$

3.2 If $y = \sin(\log(x^2 + 1))$

then $\dfrac{dy}{dx} = \dfrac{\cos(\log(x^2+1))}{x^2+1} \cdot 2x = \dfrac{2x \cos(\log(x^2+1))}{x^2+1}$

3.3 If $y = \sqrt{\sin(\log(x^2+1))}$

then $\dfrac{dy}{dx} = \dfrac{1}{2\sqrt{\sin(\log(x^2+1))}} \cdot \dfrac{\cos(\log(x^2+1))}{x^2+1} \cdot 2x$

3.4 If $y = \log(\sqrt{\sin(\log(x^2+1))}) = \log(\sin(\log(x^2+1)))^{\frac{1}{2}}$

$$= \dfrac{1}{2} \log(\sin(\log(x^2+1)))$$

then $\dfrac{dy}{dx} = \dfrac{1}{2 \sin(\log(x^2+1))} \cdot \dfrac{\cos(\log(x^2+1))}{x^2+1} \cdot 2x = \dfrac{x \cos(\log(x^2+1))}{(x^2+1)\sin(\log(x^2+1))}$

4. If $y = \dfrac{1}{\sqrt{(3e^{\sin x} - 5\cosh 4x + 1)}} = (3e^{\sin x} - 5\cosh 4x + 1)^{-\frac{1}{2}}$

then $\dfrac{dy}{dx} = -\dfrac{1}{2}(3e^{\sin x} - 5\cosh 4x + 1)^{-\frac{3}{2}}(3e^{\sin x}\cdot \cos x - 20 \sinh 4x)$

$$= -\dfrac{3e^{\sin x}\cos x - 20 \sinh 4x}{2(3e^{\sin x} - 5\cosh 4x + 1)^{\frac{3}{2}}}$$

5. If $y = e^{\operatorname{sech}^{-1}\frac{1}{x}}$, then $\dfrac{dy}{dx} = e^{\operatorname{sech}^{-1}\frac{1}{x}} \cdot -\dfrac{1}{\frac{1}{x}\sqrt{1-\left(\frac{1}{x}\right)^2}} \cdot -\dfrac{1}{x^2} = \dfrac{e^{\operatorname{sech}^{-1}\frac{1}{x}}}{\sqrt{x^2-1}}$

Exercise : Find $\dfrac{dy}{dx}$ for the following functions

1. $y = \cosh^{-1}(\cos(\cosh e^x))$ 2. $y = \tan(\sin(\operatorname{sech} x))$

3. $y = \log(\cot^3 \sqrt{x})$ 4. $y = \sqrt[3]{2x^{\frac{1}{5}} + 3\sin x - \dfrac{9}{x^{-7}}}$

5. $y = \cos^{-1}(\tan^4(\dfrac{1}{x}))$

Solutions :

1. If $y = \cosh^{-1}(\cos(\cosh e^x))$,

then $\dfrac{dy}{dx} = \dfrac{1}{\sqrt{(\cos(\cosh e^x))^2 - 1}} \cdot [-\sin(\cosh e^x) \cdot \sinh e^x \cdot e^x]$

$= -\dfrac{e^x \cdot \sinh e^x \cdot \sin(\cosh e^x)}{\sqrt{(\cos(\cosh e^x))^2 - 1}}$

2. If $y = \tan(\sin(\operatorname{sech} x))$,

then $\dfrac{dy}{dx} = -\sec^2(\sin(\operatorname{sech} x)) \cdot \cos(\operatorname{sech} x) \cdot \operatorname{sech} x \cdot \tanh x$

3. If $y = \log(\cot^3 \sqrt{x})$, then $\dfrac{dy}{dx} = \dfrac{1}{\cot^3 \sqrt{x}} \cdot 3 \cot^2 \sqrt{x} \cdot (-\operatorname{cosec}^2 \sqrt{x} \cdot \dfrac{1}{2\sqrt{x}})$

$= -\dfrac{3 \operatorname{cosec}^2 \sqrt{x}}{2\sqrt{x} \cdot \cot \sqrt{x}} = -\dfrac{3}{\sqrt{x} \cdot \sin 2\sqrt{x}}$

4. If $y = \sqrt[3]{2x^{\frac{1}{5}} + 3 \sin x - \dfrac{9}{x^{-7}}} = (2x^{\frac{1}{5}} + 3 \sin x - 9x^7)^{\frac{1}{3}}$

then $\dfrac{dy}{dx} = \dfrac{1}{3}(2x^{\frac{1}{5}} + 3 \sin x - 9x^7)^{-\frac{2}{3}} \cdot (\dfrac{2}{5} x^{-\frac{4}{5}} + 3 \cos x - 63x^6)$

5. If $y = \cos^{-1}(\tan^4(\dfrac{1}{x}))$,

then $\dfrac{dy}{dx} = -\dfrac{1}{\sqrt{1 - (\tan^4 \dfrac{1}{x})^2}} \cdot 4 \tan^3 \dfrac{1}{x} \cdot \sec^2 \dfrac{1}{x} \cdot (-\dfrac{1}{x^2}) = \dfrac{4 \tan^3 \dfrac{1}{x} \cdot \sec^2 \dfrac{1}{x}}{x^2 \sqrt{1 - (\tan^4 \dfrac{1}{x})^2}}$

Correct the errors if any :

1. If $y = e^{\sin(\tan^2 5x)}$, then $\dfrac{dy}{dx} = e^{\sin(\tan^2 5x)} \cdot \cos(\tan^2 5x) \cdot \sec^2 5x \cdot 5$.

$= 5 \cos(\tan^2 5x) \cdot \sec^2 5x \cdot e^{\sin(\tan^2 5x)}$

2. If $y = \sec^5(3x^7 - 4)$,

then $\dfrac{dy}{dx} = 5 \sec^4(3x^7 - 4) \cdot \sec(3x^7 - 4) \cdot \tan(3x^7 - 4) \cdot 21 x^6$

$= 105 x^6 \sec^5(3x^7 - 4) \cdot \tan(3x^7 - 4)$

3. If $y = \sqrt{\cot(x^2)}$, then $\dfrac{dy}{dx} = \dfrac{[-\operatorname{cosec}^2(x^2)] 2x}{\sqrt{\cot(x^2)}} = -\dfrac{2x \operatorname{cosec}^2(x^2)}{\sqrt{\cot(x^2)}}$

4. If $y = cosec^3(x^5)$, then $\dfrac{dy}{dx} = 3\, cosec^2(x^5)[-cosec(x^5).cot(x^5)]\, 5x^4$

$$= -15x^4\, cosec^3(x^5).cot(x^5)$$

5. If $y = tanh^3(tan^{-1}x)$, then $\dfrac{dy}{dx} = 3\, tanh^2(tan^{-1}x).sech^2(tan^{-1}x).\dfrac{1}{1+x^2}$

$$= \dfrac{3\, tanh^2(tan^{-1}x).sech^2(tan^{-1}x)}{1+x^2}$$

Corrected Solution

1. $\dfrac{dy}{dx} = e^{\sin(\tan^2 5x)}.\cos(\tan^2 5x).2\tan 5x.\sec^2 5x.5$

$$= 10\cos(\tan^2 5x).\tan 5x.\sec^2 5x.\,e^{\sin(\tan^2 5x)}$$

3. $\dfrac{dy}{dx} = \dfrac{[-cosec^2(x^2)]\, 2x}{2\sqrt{cot(x^2)}} = -\dfrac{x\, cosec^2(x^2)}{\sqrt{cot(x^2)}}$ 2,4,5 – No error

Do it yourself : Find $\dfrac{dy}{dx}$ for the following functions

1. $y = \sqrt{e^{\sqrt{\sin x}}}$ 2. $y = \log(\log(\log_5 x))$,

3. $y = \cos(\cosh(cos^{-1}x))$ 4. $y = \tan^{-1}(\sin^{-1}(\cos x))$

5. If $y = e^{e^{e^{\cos x}}}$

Answers :

1. $\dfrac{e^{\sqrt{\sin x}}.\cos x}{4\sqrt{\sin x}.\sqrt{e^{\sqrt{\sin x}}}}$ 2. $\dfrac{\log_5 e}{x.\log_5 x.\log(\log_5 x)}$

3. $\dfrac{\sin(\cosh(cos^{-1}x)).\sinh(cos^{-1}x)}{\sqrt{1-x^2}}$ 4. $\dfrac{-1}{1+(\sin^{-1}\cos x)^2}$

5. $-e^{e^{e^{\cos x}}}.e^{e^{\cos x}}.e^{\cos x}.\sin x$

Unit 3 : Product Rule

If $y = u \cdot v$ (u, v are functions of x) then $\dfrac{dy}{dx} = u \cdot \dfrac{dv}{dx} + v \cdot \dfrac{du}{dx}$

Examples :

1. If $y = x^3 e^x$, then

$$\dfrac{dy}{dx} = x^3 \cdot \dfrac{d(e^x)}{dx} + e^x \cdot \dfrac{d(x^3)}{dx} = x^3 e^x + e^x \cdot 3x^2 = x^3 e^x + 3x^2 e^x$$

2. If $y = \sin x \cdot \tanh x$, then

$$\dfrac{dy}{dx} = \sin x \cdot \dfrac{d(\tanh x)}{dx} + \tanh x \cdot \dfrac{d(\sin x)}{dx} = \sin x \cdot \operatorname{sech}^2 x + \operatorname{sech} x \cdot \cos x$$

3. If $y = \sin^{-1} x \cdot \log x$, then

$$\dfrac{dy}{dx} = \sin^{-1} x \cdot \dfrac{d(\log x)}{dx} + \log x \cdot \dfrac{d(\sin^{-1} x)}{dx} = \dfrac{\sin^{-1} x}{x} + \dfrac{\log x}{\sqrt{1-x^2}}$$

4. If $y = 6\sqrt{x} \cdot \cot x$, then

$$\dfrac{dy}{dx} = 6\left(\sqrt{x} \cdot \dfrac{d(\cot x)}{dx} + \cot x \cdot \dfrac{d(\sqrt{x})}{dx}\right) = 6\left(\sqrt{x} \cdot -\operatorname{cosech}^2 x + \cot x \cdot \dfrac{1}{2\sqrt{x}}\right)$$

$$= 6\left(\dfrac{\cot x}{2\sqrt{x}} - \sqrt{x} \cdot \operatorname{cosech}^2 x\right)$$

If $y = u.v.w$ (u,v,w are functions of x)

Then $\dfrac{dy}{dx} = uv \dfrac{dw}{dx} + uw \dfrac{dv}{dx} + vw \dfrac{du}{dx}$

Example : If $y = x^3 e^x \sin x$, then

$$\dfrac{dy}{dx} = x^3 e^x \dfrac{d(\sin x)}{dx} + x^3 \sin x \dfrac{d(e^x)}{dx} + e^x \sin x \dfrac{d(x^3)}{dx}$$

$$= x^3 e^x \cos x + x^3 \sin x \, e^x + e^x \sin x \cdot 3x^2$$

$$= x^3 e^x \cos x + x^3 e^x \sin x + 3x^2 e^x \sin x$$

Examples :

1. If $y = (4-x)^6 \, e^{\sqrt{x}} - 3$, then

$$\frac{dy}{dx} = (4-x)^6 \frac{d(e^{\sqrt{x}})}{dx} + e^{\sqrt{x}} \frac{d((4-x)^6)}{dx} - \frac{d(3)}{dx}$$

$$= (4-x)^6 \frac{e^{\sqrt{x}}}{2\sqrt{x}} - 6 e^{\sqrt{x}} (4-x)^5$$

2. If $y = \sin 3x . \log(\tan x) + 9 \tanh^{-1} x$, then

$$\frac{dy}{dx} = \sin 3x . \frac{d(\log \tan x)}{dx} + \log(\tan x) \frac{d(\sin 3x)}{dx} + 9. \frac{d(\tanh^{-1} x)}{dx}$$

$$= \sin 3x . \frac{\sec^2 x}{\tan x} + 3 \log(\tan x) \cos 3x + \frac{9}{1-x^2}$$

3. If $y = \cos\left(\frac{1}{x}\right) \operatorname{cosec} x^4$, then

$$\frac{dy}{dx} = \cos\left(\frac{1}{x}\right) \frac{d(\operatorname{cosec} x^4)}{dx} + \operatorname{cosec} x^4 . \frac{d(\cos(\frac{1}{x}))}{dx}$$

$$= \cos\left(\frac{1}{x}\right) . (-4x^3 . \operatorname{cosec} x^4 . \cot x^4) + \operatorname{cosec} x^4 . (-\sin\left(\frac{1}{x}\right)) . (-\frac{1}{x^2})$$

$$= -4x^3 \cos\left(\frac{1}{x}\right) \operatorname{cosec} x^4 . \cot x^4 + \frac{1}{x^2} \operatorname{cosec} x^4 . \sin\left(\frac{1}{x}\right)$$

4. If $y = \tan(\cot x) . \sec^{17}(\sin x)$, then

$$\frac{dy}{dx} = \tan(\cot x) . \frac{d(\sec^{17}(\sin x))}{dx} + \sec^{17}(\sin x) . \frac{d(\tan(\cot x))}{dx}$$

$$= 17 \tan(\cot x) . \sec^{16}(\sin x) . \sec(\sin x) . \tan(\sin x) . \cos x$$
$$+ \sec^{17}(\sin x) . \sec^2(\cot x) . (-\operatorname{cosec}^2 x)$$

$$= 17 \cos x . \tan(\cot x) . \sec^{17}(\sin x) . \tan(\sin x)$$
$$- \sec^{17}(\sin x) . \sec^2(\cot x) . \operatorname{cosec}^2 x$$

5. If $y = \sin x . \cos 2x . \tan 3x$, find $\frac{dy}{dx}$

$$\frac{dy}{dx} = \sin x . \cos 2x . \frac{d(\tan 3x)}{dx} + \sin x . \frac{d(\cos 2x)}{dx} . \tan 3x$$
$$+ \frac{d(\sin x)}{dx} . \cos 2x . \tan 3x$$

$$= 3 \sin x . \cos 2x . \sec^2 3x - 2 \sin x . \tan 3x . \sin 2x$$

$+ \cos x . \cos 2x . \tan 3x$

Exercise : Find $\dfrac{dy}{dx}$ for the following functions

1. $y = \tanh 5x . \sec^7 \sqrt{x}$
2. $y = \cos(x^4 - 1)^8 . e^{\sin(\log x)}$
3. $y = \sqrt[5]{x^2 + 6} . \sec^{-1} 3x$
4. $y = \log_a \tan x . \log(x^6 + 1)$
5. $y = 3x^4 e^x \sin(e^x - \log x)$

Solutions :

1. If $y = \tanh 5x . \sec^7 \sqrt{x}$, then

$$\dfrac{dy}{dx} = \tanh 5x . \dfrac{d(\sec^7 \sqrt{x})}{dx} + \sec^7 \sqrt{x} . \dfrac{d(\tanh 5x)}{dx}$$

$$= \tanh 5x . \dfrac{7 \sec^6 \sqrt{x} . \sec \sqrt{x} . \tan \sqrt{x}}{2\sqrt{x}} + 5 \sec^7 \sqrt{x} . \text{sech}^2 5x$$

$$= \dfrac{7 \tanh 5x . \sec^7 \sqrt{x} . \tan \sqrt{x}}{2\sqrt{x}} + 5 \sec^7 \sqrt{x} . \text{sech}^2 5x$$

2. If $y = \cos[(x^4 - 1)^8] . e^{\sin(\log x)}$, then

$$\dfrac{dy}{dx} = \cos[(x^4 - 1)^8] . \dfrac{d(e^{\sin(\log x)})}{dx} + e^{\sin(\log x)} . \dfrac{d(\cos[(x^4-1)^8])}{dx}$$

$$= \cos[(x^4 - 1)^8] . e^{\sin(\log x)} . \dfrac{\cos(\log x)}{x}$$

$$+ e^{\sin(\log x)} . [-\sin(x^4 - 1)^8 . \{8(x^4 - 1)^7 . 4x^3\}]$$

$$= \cos[(x^4 - 1)^8] . e^{\sin(\log x)} . \dfrac{\cos(\log x)}{x}$$

$$- 32 x^3 (x^4 - 1)^7 . e^{\sin(\log x)} . \sin[(x^4 - 1)^8]$$

3. If $y = \sqrt[5]{x^2 + 6} . \sec^{-1} 3x$, then

$$\dfrac{dy}{dx} = (x^2 + 6)^{\frac{1}{5}} . \dfrac{d(\sec^{-1} 3x)}{dx} + \sec^{-1} 3x . \dfrac{d((x^2+6)^{\frac{1}{5}})}{dx}$$

$$= (x^2 + 6)^{\frac{1}{5}} . \dfrac{3}{3x\sqrt{9x^2-1}} + \sec^{-1} 3x . \dfrac{1}{5}(x^2 + 6)^{\frac{-4}{5}} . 2x$$

$$= \frac{(x^2+6)^{\frac{1}{5}}}{x\sqrt{9x^2-1}} + \frac{2x(x^2+6)^{\frac{-4}{5}}}{5} \cdot \sec^{-1} 3x$$

4. If $y = \log_a \tan x \cdot \log(x^6 + 1)$, then

$$\frac{dy}{dx} = \log_a \tan x \cdot \frac{d(\log(x^6+1))}{dx} + \log(x^6 + 1) \cdot \frac{d(\log_a \tan x)}{dx}$$

$$= \log_a \tan x \cdot \frac{6x^5}{x^6+1} + \log(x^6 + 1) \cdot \frac{1}{\tan x} \sec^2 x \, \log_a e$$

$$= \frac{6x^5 \cdot \log_a \tan x}{x^6+1} + \frac{\log(x^6+1) \cdot \sec^2 x \cdot \log_a e}{\tan x}$$

5. If $y = 3x^4 e^x \sin(e^x - \log x)$, then

$$\frac{dy}{dx} = 3[x^4 e^x \cdot \cos(e^x - \log x) \cdot (e^x - \tfrac{1}{x}) + x^4 e^x \sin(e^x - \log x)$$

$$+ 4x^3 e^x \sin(e^x - \log x)]$$

$$= 3x^3 e^x [x\left(e^x - \tfrac{1}{x}\right) \cdot \cos(e^x - \log x) + x \sin(e^x - \log x)$$

$$+ 4 \sin(e^x - \log x)]$$

Correct the errors if any:

1. If $y = \cot(e^{2x}) \cdot \text{cosec}^{-1} \sqrt{x}$, then

$$\frac{dy}{dx} = \cot(e^{2x}) \cdot \frac{d(\text{cosec}^{-1}\sqrt{x})}{dx} + \text{cosec}^{-1}\sqrt{x} \cdot \frac{d(\cot(e^{2x}))}{dx}$$

$$= \cot(e^{2x}) \cdot \left(-\frac{1}{\sqrt{x}\sqrt{(\sqrt{x})^2-1}} \cdot \frac{1}{2\sqrt{x}}\right) + \text{cosec}^{-1}\sqrt{x} \cdot \text{cosec}^2(e^{2x}) \cdot (e^{2x})$$

$$= -\frac{\cot(e^{2x})}{2x\sqrt{x-1}} + e^{2x} \cdot \text{cosec}^{-1}\sqrt{x} \cdot \text{cosec}^2(e^{2x})$$

2. If $y = (3x^6 + 7x - 1)^5 \cdot \cosh\left(\tfrac{1}{x}\right)$, then

$$\frac{dy}{dx} = (3x^6 + 7x - 1)^5 \cdot \frac{d(\cosh(\tfrac{1}{x}))}{dx} + \cosh\left(\tfrac{1}{x}\right) \cdot \frac{d((3x^6+7x-1)^5)}{dx}$$

$$= (3x^6 + 7x - 1)^5 \cdot \sinh\left(\tfrac{1}{x}\right) \cdot \tfrac{1}{x^2}$$

$$+ \cosh\left(\frac{1}{x}\right) . 5(3x^6 + 7x - 1)^4(18x + 7)$$

$$= \frac{(3x^6+7x-1)^5}{x^2} . \sinh\left(\frac{1}{x}\right) + 5(3x^6 + 7x - 1)^4 . (18x + 7). \cosh\left(\frac{1}{x}\right)$$

3. $y = \log(\tan e^x) . \frac{1}{\sqrt[3]{(7x-3)}} = \log(\tan e^x) . (7x - 3)^{\frac{-1}{3}}$

$$\frac{dy}{dx} = \log(\tan e^x) . \frac{d((7x-3)^{\frac{-1}{3}})}{dx} + (7x - 3)^{\frac{-1}{3}} . \frac{d(\log(\tan e^x))}{dx}$$

$$= \log(\tan e^x) . \left[-\frac{7}{3}.(7x - 3)^{\frac{-4}{3}}\right] + (7x - 3)^{\frac{-1}{3}} . \frac{\sec^2 e^x}{\tan e^x} . e^x$$

$$= -\frac{7(7x-3)^{\frac{-4}{3}}. \log(\tan e^x)}{3} + (7x - 3)^{\frac{-1}{3}} . e^x . \frac{\sec^2 e^x}{\tan e^x}$$

4. If $y = \sec^2 x . \csc^{-1}x + \coth 4x . \log_a \sin x$, then

$$\frac{dy}{dx} = \sec^2 x . \frac{d(\csc^{-1}x)}{dx} + \csc^{-1}x . \frac{d(\sec^2 x)}{dx}$$

$$+ \coth 4x . \frac{d(\log_a \sin x)}{dx} + \log_a \sin x . \frac{d(\coth 4x)}{dx}$$

$$= \sec^2 x . \left(-\frac{1}{x\sqrt{x^2-1}}\right) + \csc^{-1}x . 2\sec^2 x. \tan x$$

$$+ \coth 4x . \frac{\cos x . \log_a e}{\sin x} + \log_a \sin x . (-4 \operatorname{cosech}^2 4x)$$

$$= -\frac{\sec^2 x}{x\sqrt{x^2-1}} + 2 \csc^{-1}x . \sec^2 x. \tan x$$

$$+ \coth 4x . \cot x . \log_a e - 4 \operatorname{cosech}^2 4x. \log_a \sin x.$$

5. If $y = \sqrt[n]{x} . e^{\cos x} . \log(\tan^{-1}x)$.

$$\frac{dy}{dx} = \sqrt[n]{x} . e^{\cos x} . \frac{1}{\tan^{-1}x} . \frac{1}{\sqrt{1+x^2}} + \sin x . e^{\cos x} . \sqrt[n]{x} . \log(\tan^{-1}x)$$

$$+ \frac{1}{n}. x^{n+1} . e^{\cos x} . \log(\tan^{-1}x)$$

CALCULUS I – Differential Calculus Formulae Practice Workbook Vol II

Corrected Solutions (CS)

1. CS: $-\dfrac{\cot(e^{2x})}{2x\sqrt{x-1}} - 2e^{2x} \cdot \operatorname{cosec}^{-1}\sqrt{x} \cdot \operatorname{cosec}^2(e^{2x})$

2. CS: $-\dfrac{(3x^6+7x-1)^5}{x^2} \cdot \sinh\left(\dfrac{1}{x}\right) + 5(3x^6+7x-1)^4 \cdot (18x^5+7) \cdot \cosh\left(\dfrac{1}{x}\right)$

5. CS: $\sqrt[n]{x} \cdot e^{\cos x} \cdot \dfrac{1}{\tan^{-1}x} \cdot \dfrac{1}{1+x^2} - \sin x \cdot e^{\cos x} \cdot \sqrt[n]{x} \cdot \log(\tan^{-1}x)$
$$+ \dfrac{1}{n} \cdot x^{\frac{1}{n}-1} \cdot e^{\cos x} \cdot \log(\tan^{-1}x)$$

3,4 – No error

Do it yourself: Find $\dfrac{dy}{dx}$ for the following functions

1. $y = 3x^4 e^x + 2\cos x - 5$
2. $y = \tan 3x \cdot \tan(x^3)$
3. $y = \sqrt{\tan^{-1}x} \cdot \sin(\log(\operatorname{cosec} x))$
4. $y = \sin x \cdot e^x \cdot \sqrt{\sqrt{x}+1}$
5. $y = (x^3+5x-1)^3 \cosh\left(\dfrac{1}{x}\right) - 2\tan(x^3) - 4$

Answers:

1. $3e^x x^3(x+4) - 2\sin x$

2. $3x^2 \cdot \tan 3x \cdot \sec^2(x^3) + 3\tan(x^3) \cdot \sec^2 3x$

3. $\dfrac{\sin(\log(\operatorname{cosec} x))}{2\sqrt{\tan^{-1}x}} \cdot \dfrac{1}{1+x^2} - \sqrt{\tan^{-1}x} \cdot \cos(\log(\operatorname{cosec} x)) \cdot \cot x$

4. $\dfrac{\sin x \cdot e^x}{4\sqrt{x}\sqrt{\sqrt{x}+1}} + \sin x \cdot \sqrt{\sqrt{x}+1} \cdot e^x + \sqrt{\sqrt{x}+1} \cdot e^x \cos x$

5. $\dfrac{-(x^3+5x-1)^3 \cdot \sinh\left(\dfrac{1}{x}\right)}{x^2}$
$$+ 3\cosh\left(\dfrac{1}{x}\right)[(x^3+5x-1)^2(3x^2+5)] - 6x^2 \sec^2(x^3)]$$

Unit 4 : Quotient Rule

If $y = \dfrac{u}{v}$ (u, v are functions of x) then $\dfrac{dy}{dx} = \dfrac{v \cdot \dfrac{du}{dx} - u \cdot \dfrac{dv}{dx}}{v^2}$

Here $\dfrac{dy}{dx} = \dfrac{Dr \cdot \dfrac{d(Nr)}{dx} - Nr \cdot \dfrac{d(Dr)}{dx}}{(Dr)^2}$ [Dr. – Denominator, Nr. – Numerator]

Examples :

1. If $y = \dfrac{\sin x}{e^x}$, then $\dfrac{dy}{dx} = \dfrac{e^x \cdot \dfrac{d(\sin x)}{dx} - \sin x \cdot \dfrac{d(e^x)}{dx}}{(e^x)^2}$

$$= \dfrac{e^x \cdot \cos x - \sin x \cdot e^x}{e^{2x}} = \dfrac{\cos x - \sin x}{e^x}$$

2. If $y = \dfrac{x^2 - 1}{x^2 + 1}$, then $\dfrac{dy}{dx} = \dfrac{(x^2+1) \cdot \dfrac{d(x^2-1)}{dx} - (x^2-1) \cdot \dfrac{d(x^2+1)}{dx}}{(x^2+1)^2}$

$$= \dfrac{2x(x^2+1) - 2x(x^2-1)}{(x^2+1)^2} = \dfrac{4x}{(x^2+1)^2}$$

3. If $y = \dfrac{\cosh 2x}{\tan x - \sin x}$,

then $\dfrac{dy}{dx} = \dfrac{(\tan x - \sin x) \cdot \dfrac{d(\cosh 2x)}{dx} - \cosh 2x \cdot \dfrac{d(\tan x - \sin x)}{dx}}{(\tan x - \sin x)^2}$

$$= \dfrac{2\sinh 2x \cdot (\tan x - \sin x) - \cosh 2x \cdot (\sec^2 x - \cos x)}{(\tan x - \sin x)^2}$$

4. If $y = \dfrac{\tan 5x - 1}{\tan 5x + 1}$,

then $\dfrac{dy}{dx} = \dfrac{(\tan 5x + 1) \cdot \dfrac{d(\tan 5x - 1)}{dx} - (\tan 5x - 1) \cdot \dfrac{d(\tan 5x + 1)}{dx}}{(\tan 5x + 1)^2}$

$$= \dfrac{5\sec^2 5x (\tan 5x + 1) - 5\sec^2 5x (\tan 5x - 1)}{(\tan 5x + 1)^2} = \dfrac{10 \sec^2 5x}{(\tan 5x + 1)^2}$$

CALCULUS I – Differential Calculus Formulae Practice Workbook Vol II

5. If $y = \dfrac{\text{sech } x}{\coth(e^x)}$, then $\dfrac{dy}{dx} = \dfrac{\coth(e^x) \cdot \dfrac{d(\text{sech } x)}{dx} - \text{sech } x \cdot \dfrac{d(\coth e^x)}{dx}}{\coth^2 e^x}$

$$= \dfrac{-\coth(e^x) \cdot \text{sech } x \cdot \tanh x + \text{sech } x \cdot \text{cosech}^2 e^x}{\coth^2(e^x)}$$

Exercise : Find $\dfrac{dy}{dx}$ for the following functions

1. $y = \dfrac{\sinh x \cdot \tan x - 5}{x^2}$ 2. $y = \dfrac{x \cdot e^x + \sin x \cdot \log x}{e^x}$ 3. $y = \dfrac{e^{\cos x}}{\cos(e^x)}$

4. $y = \dfrac{5x}{\tan^{-1} x}$ 5. $y = \sqrt{\dfrac{x+4}{x-3}}$

Solutions :

1. $y = \dfrac{\sinh x \cdot \tan x - 5}{x^2}$,

   then $\dfrac{dy}{dx} = \dfrac{x^2 \cdot \dfrac{d(\sinh x \cdot \tan x - 5)}{dx} - (\sinh x \cdot \tan x - 5) \cdot \dfrac{d(x^2)}{dx}}{x^4}$

   $= \dfrac{x^2 \cdot (\sinh x \cdot \sec^2 x + \tan x \cdot \cosh x) - 2x \cdot (\sinh x \cdot \tan x - 5)}{x^4}$

   $= \dfrac{x \cdot (\sinh x \cdot \sec^2 x + \tan x \cdot \cosh x) - 2(\sinh x \cdot \tan x - 5)}{x^3}$

2. $y = \dfrac{x \cdot e^x + \sin x \cdot \log x}{e^x}$,

   then $\dfrac{dy}{dx} = \dfrac{e^x [(x \cdot e^x + e^x) + (\dfrac{\sin x}{x} + \log x \cdot \cos x)] - (x \cdot e^x + \sin x \cdot \log x) e^x}{e^{2x}}$

   $= \dfrac{e^x + \dfrac{\sin x}{x} + \log x \cdot \cos x - \sin x \cdot \log x}{e^x}$

3. $y = \dfrac{e^{\cos x}}{\cos(e^x)}$, then $\dfrac{dy}{dx} = \dfrac{\cos(e^x) \cdot e^{\cos x}(-\sin x) - e^{\cos x} \cdot (-\sin(e^x)) \cdot e^x}{\cos^2(e^x)}$

   $= \dfrac{e^{\cos x}(e^x \cdot \sin(e^x) - \cos(e^x) \cdot \sin x)}{\cos^2(e^x)}$

4. $y = \dfrac{5x}{\tan^{-1}x}$, then $\dfrac{dy}{dx} = \dfrac{5\tan^{-1}x - \dfrac{5x}{1+x^2}}{(\tan^{-1}x)^2}$

$= \dfrac{5[(1+x^2).\tan^{-1}(x) - x]}{(1+x^2).(\tan^{-1}x)^2}$

5. $y = \sqrt{\dfrac{x+4}{x-3}}$, then $\dfrac{dy}{dx} = \dfrac{1}{2\sqrt{\dfrac{x+4}{x-3}}} \cdot \dfrac{(x-3).1 - (x+4).1}{(x-3)^2}$

$= \dfrac{\sqrt{x-3}}{2\sqrt{x+4}} \cdot \dfrac{-7}{(x-3)^2} = -\dfrac{7}{2\sqrt{x+4}.(x-3)^{\frac{3}{2}}}$

Correct the errors if any :

1. If $y = \dfrac{\sin^{-1}(\log x)}{\log(\sin^{-1}x)}$, then

$\dfrac{dy}{dx} = \dfrac{\sin^{-1}(\log x) \cdot \dfrac{1}{\sin^{-1}x} \cdot \dfrac{1}{\sqrt{1-x^2}} - \log(\sin^{-1}x) \cdot \dfrac{1}{\sqrt{1-(\log x)^2}} \cdot \dfrac{1}{x}}{(\log(\sin^{-1}x))^2}$

$= \dfrac{1}{(\log(\sin^{-1}x))^2}\left[\dfrac{\sin^{-1}(\log x)}{\sin^{-1}x . \sqrt{1-x^2}} - \dfrac{\log(\sin^{-1}x)}{x\sqrt{1-(\log x)^2}}\right]$

2. If $y = \dfrac{\tan\sqrt{x}}{\sqrt{\tan x}}$, then $\dfrac{dy}{dx} = \dfrac{\sqrt{\tan x} \cdot \dfrac{\sec^2\sqrt{x}}{2\sqrt{x}} - \tan\sqrt{x} \cdot \dfrac{\sec^2 x}{2\sqrt{\tan x}}}{\tan x}$

$= \dfrac{\tan x . \sec^2\sqrt{x} - \sqrt{x}.\tan\sqrt{x}.\sec^2 x}{4\sqrt{x}.(\tan x)^{\frac{3}{2}}}$

3. If $y = \dfrac{\sec(\log x)}{\log(\sec x)}$,

then $\dfrac{dy}{dx} = \dfrac{\log(\sec x).\sec(\log x).\tan(\log x) - \sec(\log x).\dfrac{\sec x . \tan x}{\sec x}}{(\log(\sec x))^2}$

$= \dfrac{\sec(\log x)[\log(\sec x).\tan(\log x) - \sec x . \tan x]}{(\log(\sec x))^2}$

4. If $y = \dfrac{\text{cosec}^{-1}x}{\tanh^{-1}x}$, then $\dfrac{dy}{dx} = \dfrac{\tanh^{-1}x \cdot -\dfrac{1}{x\sqrt{x^2-1}} - \text{cosec}^{-1}x \cdot \dfrac{1}{1+x^2}}{(\tanh^{-1}x)^2}$

$= -\dfrac{(1+x^2)\tanh^{-1}x + x\sqrt{x^2-1}\ \text{cosec}^{-1}x}{x(1+x^2)\cdot\sqrt{x^2-1}\cdot(\tanh^{-1}x)^2}$

Corrected Solutions

1. CS : $\dfrac{dy}{dx} = \dfrac{\log(\sin^{-1}x)\cdot\dfrac{1}{\sqrt{1-(\log x)^2}}\cdot\dfrac{1}{x} - \sin^{-1}(\log x)\cdot\dfrac{1}{\sin^{-1}x}\cdot\dfrac{1}{\sqrt{1-x^2}}}{(\log(\sin^{-1}x))^2}$

$= \dfrac{1}{(\log(\sin^{-1}x))^2}\left[\dfrac{\log(\sin^{-1}x)}{x\sqrt{1-(\log x)^2}} - \dfrac{\sin^{-1}(\log x)}{\sin^{-1}x\cdot\sqrt{1-x^2}}\right]$

3. CS : $\dfrac{dy}{dx} = \dfrac{\log(\sec x)\cdot\sec(\log x)\cdot\tan(\log x)\cdot\dfrac{1}{x} - \sec(\log x)\cdot\dfrac{\sec x\cdot\tan x}{\sec x}}{(\log(\sec x))^2}$

$= \dfrac{\sec(\log x)[\log(\sec x)\cdot\tan(\log x) - x\cdot\sec x\cdot\tan x]}{x(\log(\sec x))^2}$

4. CS : $\dfrac{dy}{dx} = \dfrac{\tanh^{-1}x \cdot -\dfrac{1}{x\sqrt{x^2-1}} - \text{cosec}^{-1}x \cdot \dfrac{1}{1-x^2}}{(\tanh^{-1}x)^2}$

$= -\dfrac{(1-x^2)\tanh^{-1}x + x\sqrt{x^2-1}\ \text{cosec}^{-1}x}{x(1-x^2)\cdot\sqrt{x^2-1}\cdot(\tanh^{-1}x)^2}$

2 - No error

Do it yourself : Find $\dfrac{dy}{dx}$ for the following functions

1. $y = \dfrac{\log(\sin x)}{\sin x}$

2. $y = \dfrac{e^{2x}\cdot\cosh x}{\cos x}$

3. $y = \dfrac{\tan(2x+5)}{\tanh 3x}$

4. $y = \dfrac{e^{\sec 2x}}{\sec(e^{2x})}$

5. $y = \dfrac{\text{cosec}^{-1}(\text{cosec}(e^x))}{\cot^{-1}\cot(e^x + e^{2x})}$

CALCULUS I – Differential Calculus Formulae Practice Workbook Vol II

Answers :

1. $\dfrac{\cos x \cdot (1 - \log(\sin x))}{\sin^2 x}$

2. $\dfrac{e^{2x}[\cos x(\sinh x + 2\cosh x) + \cosh x \cdot \sin x]}{\cos^2 x}$

3. $\dfrac{2\tanh 3x \cdot \sec^2(2x+5) - 3\tan(2x+5) \cdot \text{sech}^2\, 3x}{\tanh^2 3x}$

4. $\dfrac{2\, e^{\sec 2x} \cdot \sec(e^{2x})[\sec 2x \cdot \tan 2x - \tan(e^{2x}) \cdot e^{2x}]}{\sec^2(e^{2x})}$

5. $y = \dfrac{\text{cosec}^{-1}(\text{cosec}(e^x))}{\cot^{-1}\cot(e^x + e^{2x})} = \dfrac{e^x}{e^x(1+e^x)} = (1+e^x)^{-1}$

$\dfrac{dy}{dx} = \dfrac{-e^x}{(1+e^x)^2}$

Unit 5 Differentiation of Implicit Functions

So for we have differentiated functions of the form $y = f(x)$. Here as y is directly a function of x, these functions are called 'explicit functions'. There are functions in which y is a function of x but not directly. Example : $x + e^{xy} - 5 = 0$. Functions of this type are called 'implicit functions'.

$$\frac{d(y^2)}{dx} = 2y \cdot \frac{dy}{dx} \qquad\qquad \frac{d(\sqrt{y})}{dx} = \frac{1}{2\sqrt{y}} \cdot \frac{dy}{dx}$$

$$\frac{d(\frac{1}{y})}{dx} = -\frac{1}{y^2} \cdot \frac{dy}{dx} \qquad\qquad \frac{d(\sin y)}{dx} = \cos y \cdot \frac{dy}{dx}$$

$$\frac{d(e^y)}{dx} = e^y \cdot \frac{dy}{dx} \qquad\qquad \frac{d(\tan^{-1} y)}{dx} = \frac{1}{1+y^2} \cdot \frac{dy}{dx} \quad \ldots \text{etc.}$$

Examples:

1. $x^2 + y^2 = a^2$, Diff. w.r.to x, we get $2x + 2y \cdot \frac{dy}{dx} = 0 \implies \frac{dy}{dx} = -\frac{x}{y}$

2. $y = x \sin y$, Diff. w.r.to x, we get $\frac{dy}{dx} = x \cos y \cdot \frac{dy}{dx} + \sin y$

$$\implies \frac{dy}{dx} = \frac{\sin y}{1 - x \cos y}$$

3. $e^x + e^{-y} = e^{x-y}$, Diff. w.r.to x, we get

$$e^x - e^{-y} \cdot \frac{dy}{dx} = e^{x-y}(1 - \frac{dy}{dx}) \implies e^{x-y} \cdot \frac{dy}{dx} - e^{-y} \cdot \frac{dy}{dx} = e^{x-y} - e^x$$

$$\implies e^{-y}.(e^x - 1)\frac{dy}{dx} = e^x(e^{-y} - 1)$$

$$\implies \frac{dy}{dx} = \frac{e^x(e^{-y}-1)}{e^{-y}(e^x-1)}$$

4. $xy + y^2 = \tan x \sin y$, Diff. w.r.to x, we get

$$x \cdot \frac{dy}{dx} + y + 2y \cdot \frac{dy}{dx} = \tan x \cdot \cos y \cdot \frac{dy}{dx} + \sin y \cdot \sec^2 x$$

$$(x + 2y - \tan x \cdot \cos y) \cdot \frac{dy}{dx} = \sin y \cdot \sec^2 x - y$$

$$\Rightarrow \frac{dy}{dx} = \frac{\sin y \cdot \sec^2 x - y}{x + 2y - \tan x \cdot \cos y}$$

5. $\frac{x^2}{a^2} + \frac{y^2}{b^2} = 1$, Diff. w.r.to x, we get

$$\frac{1}{a^2} \cdot 2x + \frac{1}{b^2} \cdot 2y \cdot \frac{dy}{dx} = 0 \Rightarrow \frac{dy}{dx} = -\frac{b^2 x}{a^2 y}$$

Exercise : Find the derivative of the following w.r.to x

1. $\sin^2 y + \cos xy = k$ 2. $x^3 + y^3 = 3axy$ 3. $\left(\frac{x}{a}\right)^{\frac{1}{2}} + \left(\frac{y}{b}\right)^{\frac{1}{2}} = 1$

4. $y \tan x + y^2 = \frac{y}{x}$ 5. $7 \sinh y + \tan^{-1} y = 2x$

Solutions :

1. $\sin^2 y + \cos xy = k$, Diff. w.r.to x, we get

$$2 \sin y \cdot \cos y \cdot \frac{dy}{dx} - \sin xy \cdot \left(x \cdot \frac{dy}{dx} + y\right) = 0$$

$$(2 \sin y \cdot \cos y - x \cdot \sin xy) \frac{dy}{dx} = y \cdot \sin xy$$

$$\Rightarrow \frac{dy}{dx} = \frac{y \cdot \sin xy}{\sin 2y - x \cdot \sin xy}$$

2. $x^3 + y^3 = 3axy$, Diff. w.r.to x, we get

$$3x^2 + 3y^2 \cdot \frac{dy}{dx} = 3a\left(x \cdot \frac{dy}{dx} + y\right) \Rightarrow 3(y^2 - ax) \frac{dy}{dx} = 3(ay - x^2)$$

$$\Rightarrow \frac{dy}{dx} = \frac{ay - x^2}{y^2 - ax}$$

3. $\left(\frac{x}{a}\right)^{\frac{1}{2}} + \left(\frac{y}{b}\right)^{\frac{1}{2}} = 1 \Rightarrow \frac{\sqrt{x}}{\sqrt{a}} + \frac{\sqrt{y}}{\sqrt{b}} = 1 \Rightarrow \sqrt{b}\sqrt{x} + \sqrt{a}\sqrt{y} = \sqrt{ab}$

Diff. w.r.to x, we get, $\frac{\sqrt{b}}{2\sqrt{x}} + \frac{\sqrt{a}}{2\sqrt{y}} \cdot \frac{dy}{dx} = 0$

$$\Rightarrow \frac{dy}{dx} = -\frac{\sqrt{by}}{\sqrt{ax}}$$

CALCULUS I – Differential Calculus Formulae Practice Workbook Vol II

4. $y \tan x + y^2 = \dfrac{y}{x}$, Diff. w.r.to x, we get

$$y.\sec^2 x + \tan x.\dfrac{dy}{dx} + 2y.\dfrac{dy}{dx} = \dfrac{x.\dfrac{dy}{dx} - y}{x^2} = \dfrac{1}{x}.\dfrac{dy}{dx} - \dfrac{y}{x^2}$$

$$\Rightarrow (\tan x + 2y - \dfrac{1}{x})\dfrac{dy}{dx} = -(\dfrac{y}{x^2} + y.\sec^2 x)$$

$$\Rightarrow \dfrac{dy}{dx} = -\dfrac{y + x^2 y \sec^2 x}{x(x \tan x + 2xy - 1)}$$

5. $7 \sinh y + \tan^{-1} y = 2x$, Diff. w.r.to x, we get

$$7 \cosh y.\dfrac{dy}{dx} + \dfrac{1}{1+y^2}.\dfrac{dy}{dx} = 2 \Rightarrow (7 \cosh y + \dfrac{1}{1+y^2})\dfrac{dy}{dx} = 2$$

$$\Rightarrow \dfrac{7(1+y^2)\cosh y + 1}{1+y^2}.\dfrac{dy}{dx} = 2$$

$$\Rightarrow \dfrac{dy}{dx} = \dfrac{2(1+y^2)}{7(1+y^2)\cosh y + 1}$$

Correct the following errors if any:

1. $y \cos x = \sin y$, Diff. w.r.to x, we get

$$-y.\sin x + \cos x.\dfrac{dy}{dx} = \cos y \Rightarrow \cos x.\dfrac{dy}{dx} = \cos y + y.\sin x$$

$$\Rightarrow \dfrac{dy}{dx} = \dfrac{\cos y + y.\sin x}{\cos x}$$

2. $(1+y^2)\sec x + y \cot x = 3 \log y$, Diff. w.r.to x, we get

$$(1+y^2)\sec x.\tan x + \sec x.2y.\dfrac{dy}{dx} - y \csc^2 x + \cot x.\dfrac{dy}{dx} = \dfrac{3}{y}.\dfrac{dy}{dx}$$

$$\Rightarrow (2y.\sec x + \cot x - \dfrac{3}{y}).\dfrac{dy}{dx} = y \csc^2 x - (1+y^2)\sec x.\tan x$$

$$\Rightarrow \dfrac{dy}{dx} = \dfrac{y^2 \csc^2 x - y(1+y^2)\sec x.\tan x}{2y^2.\sec x + y.\cot x - 3}$$

3. $\sin(x+y) + \cos(x-y) = 1$, Diff. w.r.to x, we get

$$\cos(x+y).(1 + \dfrac{dy}{dx}) - \sin(x-y).(1 - \dfrac{dy}{dx}) = 0$$

$$\Rightarrow [\cos(x+y) + \sin(x-y)].\frac{dy}{dx} = \sin(x-y) - \cos(x+y)$$

$$\Rightarrow \frac{dy}{dx} = \frac{\sin(x-y) - \cos(x+y)}{\cos(x+y) + \sin(x-y)}$$

4. $x\,\tan^{-1}y - y\,\tan^{-1}x = 1$, Diff. w.r.to x, we get

$$x.\frac{1}{1+y^2}.\frac{dy}{dx} + \tan^{-1}y - y.\frac{1}{1+x^2} - \tan^{-1}x.\frac{dy}{dx} = 0$$

$$\Rightarrow \left(\frac{x}{1+y^2} - \tan^{-1}x\right).\frac{dy}{dx} = \frac{y}{1+y^2} - \tan^{-1}y$$

$$\Rightarrow \frac{x - (1+y^2)\tan^{-1}x}{1+y^2}.\frac{dy}{dx} = \frac{y - (1+y^2)\tan^{-1}y}{1+y^2}$$

$$\Rightarrow \frac{dy}{dx} = \frac{y - (1+y^2)\tan^{-1}y}{x - (1+y^2)\tan^{-1}x}$$

5. $\cosh x.\sinh y = \operatorname{sech} y.\coth x$, Diff. w.r.to x, we get

$$\sinh x.\cosh y.\frac{dy}{dx} = \operatorname{sech} y.(-\operatorname{cosech}^2 x) + \coth x.(-\operatorname{sech} y.\tanh y).\frac{dy}{dx}$$

$$\Rightarrow (\sinh x.\cosh y + \coth x.\operatorname{sech} y.\tanh y)\frac{dy}{dx} = \operatorname{sech} y.\operatorname{cosech}^2 x$$

$$\Rightarrow \frac{dy}{dx} = \frac{\operatorname{sech} y.\operatorname{cosech}^2 x}{\sinh x.\cosh y + \coth x.\operatorname{sech} y.\tanh y}$$

Corrected Solution

1. CS : $-y.\sin x + \cos x.\dfrac{dy}{dx} = \cos y.\dfrac{dy}{dx}$

$$\Rightarrow (\cos x - \cos y)\frac{dy}{dx} = y.\sin x$$

$$\Rightarrow \frac{dy}{dx} = \frac{y.\sin x}{\cos x - \cos y}$$

5. CS : $\cosh x.\cosh y.\dfrac{dy}{dx} + \sinh y.\sinh x$

$$= \operatorname{sech} y.(-\operatorname{cosech}^2 x) + \coth x.(-\operatorname{sech} y.\tanh y).\frac{dy}{dx}$$

$$\Rightarrow (\cosh x.\cosh y + \coth x.\operatorname{sech} y.\tanh y)\frac{dy}{dx}$$

$$= -(\operatorname{sech} y.\operatorname{cosech}^2 x + \sinh y.\sinh x)$$

$$\Rightarrow \quad \frac{dy}{dx} = -\frac{\operatorname{sech} y \cdot \operatorname{cosech}^2 x + \sinh y \cdot \sinh x}{\cosh x \cdot \cosh y + \coth x \cdot \operatorname{sech} y \cdot \tanh y}$$

2,3,4 – No error

Do it yourself :

1. $xy = 3(x + y)$
2. $\dfrac{x \cdot \sin x}{y} = \log y$
3. $xe^{-y} + ye^x = y^2$

4. $\log(\sec y) = \sec(\log y) + x^2$

5. $ax^2 + by^2 + 2gx + 2fy + 2hxy + c = 0$

Answers :

1. $\dfrac{3-y}{x-3}$
2. $\dfrac{x \cdot \cos x + \sin x}{1 + \log y}$
3. $-\dfrac{ye^x + e^{-y}}{e^x - x e^{-y} - 2y}$

4. $\dfrac{2xy}{y \tan y - \sec(\log y) \cdot \tan(\log y)}$
5. $-\dfrac{ax + hy + g}{hx + by + f}$

Unit 6 Parametric Form

A relation expressed between two variables x and y in the form $x = f(t)$, $y = f(t)$ is said to be in 'parametric form' with the parameter t. Here $\dfrac{dy}{dx} = \dfrac{dy/dt}{dx/dt}$

Try to remember:

Derivative of y w.r.to x is $\dfrac{dy}{dx}$ Derivative of y w.r.to y is $\dfrac{dy}{dy} = 1$

Derivative of x w.r.to y is $\dfrac{dx}{dy}$ Derivative of x w.r.to x is $\dfrac{dx}{dx} = 1$

Derivative of y w.r.to t is $\dfrac{dy}{dt}$ Derivative of t w.r.to t is $\dfrac{dt}{dt} = 1$

Derivative of x w.r.to t is $\dfrac{dx}{dt}$ Derivative of θ w.r.to θ is $\dfrac{d\theta}{d\theta} = 1$

Derivative of y w.r.to θ is $\dfrac{dy}{d\theta}$ Derivative of x w.r.to θ is $\dfrac{dx}{d\theta}$ etc.

Examples :

1. $x = at^2, y = 2at \Rightarrow \dfrac{dx}{dt} = 2at \ ; \ \dfrac{dy}{dt} = 2a \Rightarrow \dfrac{dy}{dx} = \dfrac{1}{t}$

2. $x = a\cos^3 t, y = a\sin^3 t \Rightarrow \dfrac{dx}{dt} = -3a\cos^2 t.\sin t \ ; \ \dfrac{dy}{dt} = 3a\sin^2 t.\cos t$

$\Rightarrow \dfrac{dy}{dx} = -\dfrac{3a\sin^2 t.\cos t}{3a\cos^2 t.\sin t} = -\dfrac{\sin t}{\cos t} = -\tan t$

3. $x = 4t, \ y = \dfrac{4}{t} \Rightarrow \dfrac{dx}{dt} = 4 \ ; \ \dfrac{dy}{dt} = -\dfrac{4}{t^2} \Rightarrow \dfrac{dy}{dx} = -\dfrac{4/t^2}{4} = -\dfrac{1}{t^2}$

4. $x = a(\cos t + t.\sin t), y = a(\sin t - t.\cos t)$

$\Rightarrow \dfrac{dx}{dt} = a(-\sin t + t.\cos t + \sin t) = a\,t\cos t$

$\dfrac{dy}{dt} = a[\cos t - (-t\sin t + \cos t)] = a\,t\sin t \Rightarrow \dfrac{dy}{dx} = \tan t$

5. $x = a(\log \cot \dfrac{\theta}{2} - \cos \theta), y = a\sin\theta$

$$\Rightarrow \frac{dx}{d\theta} = a[\frac{1}{\cot\frac{\theta}{2}}(-\csc^2\frac{\theta}{2}).\frac{1}{2} + \sin\theta]$$

$$= a[-\frac{\sin\frac{\theta}{2}}{2\sin^2\frac{\theta}{2}.\cos\frac{\theta}{2}} + \sin\theta] = a(-\frac{1}{\sin\theta} + \sin\theta) = -a\frac{\cos^2\theta}{\sin\theta}$$

$$\frac{dy}{d\theta} = a\cos\theta \Rightarrow \frac{dy}{dx} = -\frac{a\cos\theta.\sin\theta}{a\cos^2\theta} = -\tan\theta$$

Exercise :

1. $x = \sin t, y = \cos 2t$
2. $x = a(\theta - \sin\theta), y = a(1 + \cos\theta)$

3. $x = a(\cos t + \log\tan\frac{t}{2}), y = a\sin t$
4. $x = e^t \cos t, y = e^t \sin t$

5. $x = a\log(\sec\theta), y = a(\tan\theta - \theta)$

1. $x = \sin t, y = \cos 2t \Rightarrow \frac{dx}{dt} = \cos t ; \frac{dy}{dt} = -2\sin 2t$

$$\Rightarrow \frac{dy}{dx} = -\frac{2\sin 2t}{\cos t} = -4\sin t$$

2. $x = a(\theta - \sin\theta), y = a(1 + \cos\theta)$

$$\Rightarrow \frac{dx}{d\theta} = a(1 - \cos\theta) ; \frac{dy}{d\theta} = -a\sin\theta$$

$$\Rightarrow \frac{dy}{dx} = -\frac{a\sin\theta}{a(1 - \cos\theta)} = -\frac{2\sin\frac{\theta}{2}.\cos\frac{\theta}{2}}{2\sin^2\frac{\theta}{2}} = -\cot\frac{\theta}{2}$$

3. $x = a(\cos t + \log\tan\frac{t}{2}), y = a\sin t$

$$\Rightarrow \frac{dx}{dt} = a(-\sin t + \frac{1}{\tan\frac{t}{2}}.\sec^2\frac{t}{2}.\frac{1}{2}) = a(-\sin t + \frac{1}{2\sin\frac{t}{2}.\cos\frac{t}{2}})$$

$$= a(-\sin t + \frac{1}{\sin t}) = a(\frac{1 - \sin^2 t}{\sin t}) = a\frac{\cos^2 t}{\sin t}$$

$$\frac{dy}{dt} = a\cos t \Rightarrow \frac{dy}{dx} = \tan t$$

4. $x = e^t \cos t, y = e^t \sin t \Rightarrow \dfrac{dx}{dt} = -e^t \sin t + e^t \cos t = e^t(\cos t - \sin t)$

$$\dfrac{dy}{dt} = e^t \cos t + e^t \sin t = e^t(\cos t + \sin t)$$

$\Rightarrow \dfrac{dy}{dx} = \dfrac{\cos t + \sin t}{\cos t - \sin t}$

5. $x = a \log(\sec \theta), y = a(\tan \theta - \theta)$

$\Rightarrow \dfrac{dx}{d\theta} = a \dfrac{\sec \theta . \tan \theta}{\sec \theta} = a \tan \theta \;;\; \dfrac{dy}{d\theta} = a(\sec^2 \theta - 1) = a \tan^2 \theta$

$\Rightarrow \dfrac{dy}{dx} = \tan \theta$

Correct the errors if any :

1. $x = ae^t(\sin t - \cos t), y = ae^t(\sin t + \cos t)$

$\Rightarrow \dfrac{dx}{dt} = ae^t(\cos t + \sin t) + ae^t(\sin t - \cos t) = 2ae^t . \sin t$

$\dfrac{dy}{dt} = ae^t(\cos t - \sin t) + ae^t(\sin t + \cos t) = 2ae^t . \cos t$

$\Rightarrow \dfrac{dy}{dx} = \tan t$

2. $x = a \log(\sec \theta + \tan \theta), y = a \sec \theta$

$\Rightarrow \dfrac{dx}{d\theta} = a \dfrac{\sec \theta . \tan \theta + \sec^2 \theta}{\sec \theta + \tan \theta} = a \sec \theta \;;\; \dfrac{dy}{d\theta} = a \sec \theta . \tan \theta$

$\Rightarrow \dfrac{dy}{dx} = \tan \theta$

3. $x^{\frac{2}{3}} + y^{\frac{2}{3}} = a^{\frac{2}{3}}$, Substituting, $x = a \cos^3 t, y = a \sin^3 t$ in $x^{\frac{2}{3}} + y^{\frac{2}{3}}$, we get

$$x^{\frac{2}{3}} + y^{\frac{2}{3}} = a^{\frac{2}{3}}(\cos^2 t + \sin^2 t) = a^{\frac{2}{3}}$$

This shows that $x = a \cos^3 t, y = a \sin^3 t$ is the parametric form of $x^{\frac{2}{3}} + y^{\frac{2}{3}} = a^{\frac{2}{3}}$

CALCULUS I – Differential Calculus Formulae Practice Workbook Vol II

$x = a\cos^3 t, y = a\sin^3 t \Rightarrow \dfrac{dx}{dt} = -3a\cos^2 t \cdot \sin t \;;\; \dfrac{dy}{dt} = 3a\sin^2 t \cdot \cos t$

$$\Rightarrow \dfrac{dy}{dx} = -\dfrac{3a\sin^2 t \cdot \cos t}{3a\cos^2 t \cdot \sin t} = -\dfrac{\sin t}{\cos t} = -\tan t$$

4. $x = \cos\theta - \cos 2\theta,\; y = \sin\theta - \sin 2\theta$

$$\Rightarrow \dfrac{dx}{d\theta} = -\sin\theta + 2\sin 2\theta \;;\; \dfrac{dy}{d\theta} = \cos\theta - 2\cos 2\theta$$

$$\Rightarrow \dfrac{dy}{dx} = \dfrac{2\sin 2\theta - \sin\theta}{\cos\theta - 2\cos 2\theta}$$

5. $x = \dfrac{3at}{1+t^3},\; y = \dfrac{3at^2}{1+t^3}$

$$\Rightarrow \dfrac{dx}{dt} = \dfrac{3a(1+t^3) - 9at^3}{(1+t^3)^2} = \dfrac{3a(1-2t^3)}{(1+t^3)^2};$$

$$\dfrac{dy}{dt} = \dfrac{6at(1+t^3) - 9at^4}{(1+t^3)^2} = \dfrac{3at(2-t^3)}{(1+t^3)^2} \Rightarrow \dfrac{dy}{dx} = \dfrac{t(2-t^3)}{1-2t^3}$$

Corrected Solution

1. CS : $\dfrac{dy}{dx} = \cot t$

4. CS : $\dfrac{dy}{dx} = \dfrac{\cos\theta - 2\cos 2\theta}{2\sin 2\theta - \sin\theta}$

2, 3, 5 – No error

Do it yourself : Find the derivative of the following

1. $x = a\sec\theta,\; y = b\tan\theta$
2. $x = \dfrac{a}{3}(t - t^3),\; y = at^2$

3. $x = a\log\tan\left(\dfrac{\pi}{4} + \dfrac{\theta}{2}\right),\; y = a\sec\theta$
4. $x = \sec e^t,\; y = \tan e^t$

5. $x = a(\cos\theta + \theta\sin\theta),\; y = a(\sin\theta - \theta\cos\theta)$

Answers :

1. $\dfrac{b}{a}\operatorname{cosec}\theta$ 2. $\dfrac{6t}{1-3t^2}$ 3. $\tan\theta$ 4. $\operatorname{cosec} e^t$ 5. $\tan\theta$

CALCULUS I – Differential Calculus Formulae Practice Workbook Vol II

Unit 7 Substitution method :

This is an alternate method to the chain rule (function of a function). It helps us to differentiate some functions which are difficult in nature..

Examples :

1. $y = sin^{-1}(3x - 4x^3)$; Substitute $x = sin\ \theta \implies \theta = sin^{-1}x$

   then $y = sin^{-1}(3sin\ \theta - 4\ sin^3\ \theta) = sin^{-1}(sin\ 3\theta) = 3\theta = 3sin^{-1}x$

   Differentiating w. r. to x, $\dfrac{dy}{dx} = \dfrac{3}{\sqrt{1-x^2}}$ (How is it?)

2. $y = cos^{-1}\dfrac{a^2 - x^2}{a^2 + x^2}$; Substitute $x = a\ tan\ \theta \implies \theta = tan^{-1}\dfrac{x}{a}$

   then $y = cos^{-1}\dfrac{a^2 - a^2 tan^2\ \theta}{a^2 + a^2 tan^2\ \theta} = cos^{-1}\dfrac{1 - tan^2\ \theta}{1 + tan^2\ \theta}$

   $= cos^{-1}(cos\ 2\theta) = 2\theta = 2\ tan^{-1}\dfrac{x}{a}$

   Differentiating w. r. to x, $\dfrac{dy}{dx} = \dfrac{2}{1 + (\frac{x}{a})^2} \cdot \dfrac{1}{a} = \dfrac{2a}{a^2 + x^2}$

Note : In this problem, by following this method, we have avoided the laborious chain rule method.

3. $y = cos^{-1}\sqrt{\dfrac{1 + cos\ x}{2}}$;

   Then $y = cos^{-1}\sqrt{\dfrac{1 + cos\ x}{2}} = cos^{-1}\sqrt{\dfrac{2\ cos^2\frac{x}{2}}{2}}$

   $= cos^{-1}(cos\ \dfrac{x}{2}) = \dfrac{x}{2}$

   Differentiating w. r. to x, $\dfrac{dy}{dx} = \dfrac{1}{2}$

4. $y = tan^{-1}\dfrac{3x - x^3}{1 - 3x^2}$; Substitute $x = tan\ \theta \implies \theta = tan^{-1}x$

   then $y = tan^{-1}\dfrac{3\ tan\ \theta - tan^3\ \theta}{1 - 3\ tan^2\ \theta} = tan^{-1}(tan\ 3\theta) = 3\theta = 3\ tan^{-1}x$

Differentiating w. r. to x, $\dfrac{dy}{dx} = \dfrac{3}{1+x^2}$

5. $y = \sec^{-1}\dfrac{1}{2x^2-1}$;   Substitute $x = \cos\theta \implies \theta = \cos^{-1}x$

   Then $y = \sec^{-1}\dfrac{1}{2\cos^2\theta - 1} = \sec^{-1}\dfrac{1}{\cos 2\theta}$

   $= \sec^{-1}\sec 2\theta = 2\theta = 2\cos^{-1}x$

   Differentiating w. r. to x, $\dfrac{dy}{dx} = -\dfrac{2}{\sqrt{1-x^2}}$

Exercise : Differentiate the following w.r.to x

1. $\sin^{-1}(2x\sqrt{1-x^2})$
2. $\cos^{-1}\dfrac{2x}{1+x^2}$
3. $\tan^{-1}\sqrt{\dfrac{1-\cos x}{1+\cos x}}$

4. $\cot^{-1}\dfrac{\sqrt{1+\sin x} + \sqrt{1-\sin x}}{\sqrt{1+\sin x} - \sqrt{1-\sin x}}$
5. $\tan^{-1}\dfrac{\sqrt{1+x^2}-1}{x}$

Solutions :

1. $y = \sin^{-1}(2x\sqrt{1-x^2})$;   Substitute $x = \sin\theta \implies \theta = \sin^{-1}x$

   then $y = \sin^{-1}(2\sin\theta\sqrt{1-\sin^2\theta}) = \sin^{-1}(2\sin\theta.\cos\theta)$

   $= \sin^{-1}(\sin 2\theta) = 2\theta = 2\sin^{-1}x$

   Differentiating w. r. to x, $\dfrac{dy}{dx} = \dfrac{2}{\sqrt{1-x^2}}$

2. $y = \cos^{-1}\dfrac{2x}{1+x^2}$;   Substitute $x = \tan\theta \implies \theta = \tan^{-1}x$

   then $y = \cos^{-1}\dfrac{2\tan\theta}{1+\tan^2\theta} = \cos^{-1}\dfrac{2\tan\theta}{\sec^2\theta} = \cos^{-1}(\sin 2\theta)$

   $= \cos^{-1}[\cos(\tfrac{\pi}{2} - 2\theta)] = \tfrac{\pi}{2} - 2\theta = \tfrac{\pi}{2} - 2\tan^{-1}x$

   Differentiating w. r. to x, $\dfrac{dy}{dx} = -\dfrac{2}{1+x^2} = -\dfrac{2}{1+x^2}$

3. $y = \tan^{-1}\sqrt{\dfrac{1-\cos x}{1+\cos x}}$;   then

CALCULUS I – Differential Calculus Formulae Practice Workbook Vol II

$$y = tan^{-1}\sqrt{\frac{2\sin^2\frac{x}{2}}{2\cos^2\frac{x}{2}}} = tan^{-1}(tan\frac{x}{2}) = \frac{x}{2}$$

Differentiating w. r. to x, $\dfrac{dy}{dx} = \dfrac{1}{2}$

4. $y = cot^{-1}\dfrac{\sqrt{1+\sin x} + \sqrt{1-\sin x}}{\sqrt{1+\sin x} - \sqrt{1-\sin x}}$

$$\sqrt{1+\sin x} = \sqrt{\sin^2\frac{x}{2} + \cos^2\frac{x}{2} + 2\sin\frac{x}{2}.\cos\frac{x}{2}} = \sin\frac{x}{2} + \cos\frac{x}{2}$$

Similarly, $\sqrt{1-\sin x} = \sin\frac{x}{2} - \cos\frac{x}{2}$

then $y = cot^{-1}\dfrac{\sin\frac{x}{2} + \cos\frac{x}{2} + \sin\frac{x}{2} - \cos\frac{x}{2}}{\sin\frac{x}{2} + \cos\frac{x}{2} - \sin\frac{x}{2} + \cos\frac{x}{2}} = cot^{-1}(tan\frac{x}{2})$

$$= cot^{-1}[cot(\frac{\pi}{2} - \frac{x}{2})] = \frac{\pi}{2} - \frac{x}{2}$$

Differentiating w. r. to x, $\dfrac{dy}{dx} = -\dfrac{1}{2}$

5. $y = tan^{-1}\dfrac{\sqrt{1+x^2}-1}{x}$; Substitute $x = tan\,\theta \Rightarrow \theta = tan^{-1}x$

then $y = tan^{-1}\dfrac{\sqrt{1+tan^2\theta}-1}{tan\,\theta} = tan^{-1}\dfrac{\sec\theta - 1}{tan\,\theta}$

$$= tan^{-1}\dfrac{1-\cos\theta}{\sin\theta} = tan^{-1}\dfrac{2\sin^2\frac{\theta}{2}}{2\sin\frac{\theta}{2}.\cos\frac{\theta}{2}}$$

$$= tan^{-1}(tan\frac{\theta}{2}) = \frac{\theta}{2} = \frac{1}{2}.tan^{-1}x$$

Differentiating w. r. to x, $\dfrac{dy}{dx} = \dfrac{1}{2(1+x^2)}$

Correct the errors if any :

1. $y = tan^{-1}\dfrac{\cos x + \sin x}{\cos x - \sin x}$

As, $\dfrac{\cos x + \sin x}{\cos x - \sin x} = \dfrac{1 + \tan x}{1 - \tan x}$ [By dividing the Nr. and Dr. by $\cos x$]

$$= \dfrac{1 + \tan x}{1 - 1.\tan x} = \dfrac{\tan\frac{\pi}{4} + \tan x}{1 - \tan\frac{\pi}{4}.\tan x} = \tan(\frac{\pi}{4} + x)$$

We have $y = tan^{-1}\dfrac{\cos x + \sin x}{\cos x - \sin x} = tan^{-1}[\tan\left(\frac{\pi}{4} + x\right)] = \frac{\pi}{4} + x$

Differentiating w. r. to x, $\dfrac{dy}{dx} = 1$

2. $y = tan^{-1}\dfrac{\sqrt{x} + \sqrt{a}}{1 - \sqrt{ax}}$; Differentiating w. r. to x,

$$\dfrac{dy}{dx} = \dfrac{1}{1 - \left(\dfrac{\sqrt{x} + \sqrt{a}}{1 - \sqrt{ax}}\right)^2} \cdot \dfrac{\frac{1}{2\sqrt{x}}.(1 - \sqrt{ax}) + (\sqrt{x} + \sqrt{a}).\frac{a}{2\sqrt{ax}}}{(1 - \sqrt{ax})^2}$$

....

....

[Note : This method seems to be lengthy. Isn't it? Why not we follow the following simple method]

We know that, $tan^{-1}x + tan^{-1}y = tan^{-1}\dfrac{x + y}{1 - xy}$

Hence, $y = tan^{-1}\dfrac{\sqrt{x} + \sqrt{a}}{1 - \sqrt{ax}} = tan^{-1}\sqrt{x} + tan^{-1}\sqrt{a}$

Differentiating w. r. to x, $\dfrac{dy}{dx} = \dfrac{1}{1 + (\sqrt{x})^2} \cdot \dfrac{1}{2\sqrt{x}} = \dfrac{1}{2\sqrt{x}(1+x)}$

3. $y = sin^{-1}\dfrac{2x}{1 + x^2}$; Substitute $x = \tan\theta \Rightarrow \theta = tan^{-1}x$

then $y = \sin\dfrac{2\tan\theta}{1 + \tan^2\theta} = sin^{-1}(\sin 2\theta) = 2\theta = 2\,tan^{-1}x$

,,,Differentiating w. r. to x, $\dfrac{dy}{dx} = \dfrac{2}{1 + x^2}$

4. $y = cos^{-1}\dfrac{1 - x^2}{1 + x^2}$; Substitute $x = \tan\theta \Rightarrow \theta = tan^{-1}x$

then $y = \cos^{-1}\dfrac{1-\tan^2\theta}{1+\tan^2\theta} = \cos^{-1}(\cos 2\theta) = 2\theta = 2\tan^{-1}x$

Differentiating w. r. to x, $\dfrac{dy}{dx} = \dfrac{2}{1+x^2}$

5. $y = \sin^{-1}\dfrac{1-x^2}{1+x^2}$; Substitute $x = \tan\theta \Rightarrow \theta = \tan^{-1}x$

then $y = \sin^{-1}\dfrac{1-\tan^2\theta}{1+\tan^2\theta} = \sin^{-1}(\cos 2\theta)$

$= \sin^{-1}[\sin(\frac{\pi}{2}-2\theta)] = \frac{\pi}{2} - 2\theta = \frac{\pi}{2} - 2\tan^{-1}x$

Differentiating w. r. to x, $\dfrac{dy}{dx} = -\dfrac{2}{1+x^2}$

Note : 1, 2, 3, 4, 5 – No error

Do it yourself : Find the derivative of the following w.r.to x

1. $y = \cos^{-1}(4x^3 - 3x)$

2. $y = \tan^{-1}\dfrac{\sqrt{1+x}-\sqrt{1-x}}{\sqrt{1+x}+\sqrt{1-x}}$ (substitute $x = \cos 2\theta$)

3. $y = \sin^{-1}\sqrt{\dfrac{1-\cos 2x}{2}}$ 4. $y = \sin^{-1}\dfrac{1-x^2}{1+x^2}$

5. $y = \cot^{-1}\dfrac{3x-x^3}{1-3x^2}$

Answers :

1. $-\dfrac{3}{\sqrt{1-x^2}}$ 2. $\dfrac{1}{2\sqrt{1-x^2}}$ 3. 1 4. $-\dfrac{2}{1+x^2}$ 5. $-\dfrac{3}{1+x^2}$

Unit 8 Logarithmic Differentiation

We are familiar with the formula $\dfrac{d(x^n)}{dx} = nx^{n-1}$ for any real n. In the power of x, instead of any real n, if we have a function of x then the above formula fails.

Example : 1. Find $\dfrac{d(x^{\sin x})}{dx}$ 2. If $x^y = y^x$, find $\dfrac{dy}{dx}$

Method : By taking logarithm to the base e, these problems can be solved easily.

Required formulae : (logarithm to the base e only)

1. $\log MN = \log M + \log N$; $\log MNP = \log M + \log N + \log P$

2. $\log \dfrac{M}{N} = \log M - \log N$ 3. $\log M^P = P \log M$

Note : Formula 1 and 2 changes multiplication into addition.

Formula 3 changes division into subtraction.

Formula 4 changes power into multiplication.

Examples :

1. $y = x^{\tan x}$; Taking logarithm to the base e both sides,

$\log y = \log x^{\tan x} \Rightarrow \log y = \tan x . \log x$ (formula 3)

Differentiating w.r.to x, $\dfrac{1}{y}.\dfrac{dy}{dx} = \tan x . \dfrac{1}{x} + \log x . \sec^2 x$

$\Rightarrow \dfrac{dy}{dx} = y(\tan x . \dfrac{1}{x} + \log x . \sec^2 x) = x^{\tan x - 1}(\tan x + x. \log x . \sec^2 x)$

2. $y = (\sin x)^{\cos x}$; Taking logarithm to the base e both sides,

$\log y = \log (\sin x)^{\cos x} \Rightarrow \log y = \cos x . \log (\sin x)$

Differentiating w.r.to x, $\dfrac{1}{y}.\dfrac{dy}{dx} = \cos x . \dfrac{\cos x}{\sin x} - \log (\sin x) . \sin x$

$\Rightarrow \dfrac{dy}{dx} = y(\cos x . \cot x - \sin x . \log (\sin x))$

$$= (\sin x)^{\cos x}(\cos x \cdot \cot x - \sin x \cdot \log(\sin x))$$

3. $y = \sin x \cdot \sin 2x \cdot \sin 3x$; Taking logarithm to the base e both sides,

$\log y = \log(\sin x \cdot \sin 2x \cdot \sin 3x) \Rightarrow \log y = \log \sin x + \log \sin 2x + \log \sin 3x$

Differentiating w.r.to x, $\quad \dfrac{1}{y} \cdot \dfrac{dy}{dx} = \dfrac{\cos x}{\sin x} + \dfrac{2\cos 2x}{\sin 2x} + \dfrac{3\cos 3x}{\sin 3x}$

$$\Rightarrow \dfrac{dy}{dx} = y(\cot x + 2\cot 2x + 3\cot 3x)$$

$$= \sin x \cdot \sin 2x \cdot \sin 3x \cdot (\cot x + 2\cot 2x + 3\cot 3x)$$

4. $y = \sqrt{\sin x + \sqrt{\sin x + \sqrt{\sin x + \cdots \infty}}}$

$y = \sqrt{\sin x + \sqrt{\sin x + \sqrt{\sin x + \cdots \infty}}} = \sqrt{\sin x + y}$

Taking log to the base e, $\log y = \log(\sin x + y)^{\frac{1}{2}} = \dfrac{1}{2}\log(\sin x + y)$

Differentiating w.r.to x, $\quad \dfrac{1}{y} \cdot \dfrac{dy}{dx} = \dfrac{1}{2(\sin x + y)} \cdot \left(\cos x + \dfrac{dy}{dx}\right)$

$\left[\dfrac{1}{y} - \dfrac{1}{2(\sin x + y)}\right] \cdot \dfrac{dy}{dx} = \dfrac{\cos x}{2(\sin x + y)}$

$\Rightarrow \left[\dfrac{2\sin x + y}{2y(\sin x + y)}\right] \cdot \dfrac{dy}{dx} = \dfrac{\cos x}{2(\sin x + y)} \quad \Rightarrow \dfrac{dy}{dx} = \dfrac{y \cdot \cos x}{2\sin x + y}$

5. $y = \sqrt[3]{\dfrac{(x^2+1)(x+3)}{(2x^3-1)(x+5)}},\quad$ Taking logarithm to the base e both sides,

$\log y = \log\left(\dfrac{(x^2+1)(x+3)}{(2x^3-1)(x+5)}\right)^{1/3}$

$\Rightarrow \log y = \dfrac{1}{3}[\log(x^2+1) + \log(x+3) - \log(2x^3-1) - \log(x+5)]$

Differentiating w.r.to x,

$\dfrac{1}{y} \cdot \dfrac{dy}{dx} = \dfrac{1}{3}\left[\dfrac{2x}{x^2+1} + \dfrac{1}{x+3} - \dfrac{6x^2}{2x^3-1} - \dfrac{1}{x+5}\right]$

$$\Rightarrow \frac{dy}{dx} = \frac{1}{3} \cdot \sqrt[3]{\frac{(x^2+1)(x+3)}{(2x^3-1)(x+5)}} \left[\frac{2x}{x^2+1} + \frac{1}{x+3} - \frac{6x^2}{2x^3-1} - \frac{1}{x+5}\right]$$

Exercise : Differentiate the following w.r.to x

1. $x^y = y^x$ 2. $a^{\tan e^x}$ 3. x^{x^3} 4. $(\log x)^{\cosh^{-1} x}$ 5. $\frac{(3x^2-1)(x-5)}{(x^3+1)(2x+1)}$

Solutions :

1. $x^y = y^x$; Taking logarithm to the base e both sides,

$\log x^y = \log y^x \Rightarrow y \log x = x \log y$

Differentiating w.r.to x, $\frac{y}{x} + \log x \cdot \frac{dy}{dx} = \frac{x}{y} \cdot \frac{dy}{dx} + \log y$

$$\Rightarrow \frac{dy}{dx}\left(\log x - \frac{x}{y}\right) = \log y - \frac{y}{x} \Rightarrow \frac{dy}{dx} = \frac{xy \log y - y^2}{xy \log x - x^2}$$

2. $y = a^{\tan e^x}$; Taking logarithm to the base e both sides,

$\log y = \log a^{\tan e^x} \Rightarrow \log y = \tan e^x \cdot \log a$

Differentiating w.r.to x, $\frac{1}{y} \cdot \frac{dy}{dx} = \log a \cdot e^x \cdot \sec^2 e^x$

$$\frac{dy}{dx} = a^{\tan e^x} \cdot e^x \cdot \sec^2 e^x \cdot \log a$$

3. $y = x^{x^3}$; Taking logarithm to the base e both sides,

$\log y = \log x^{x^3} \Rightarrow \log y = x^3 \cdot \log x$

Differentiating w.r.to x, $\frac{1}{y} \cdot \frac{dy}{dx} = x^3 \cdot \frac{1}{x} + 3x^2 \cdot \log x$

$$\frac{dy}{dx} = x^2 (1 + 3 \log x) \cdot x^{x^3}$$

4. $y = (\log x)^{\cosh^{-1} x}$; Taking logarithm to the base e both sides,

$\log y = \log[(\log x)^{\cosh^{-1} x}] \Rightarrow \log y = \cosh^{-1} x \cdot \log (\log x)$

Differentiating w.r.to x, $\dfrac{1}{y} \cdot \dfrac{dy}{dx} = \cosh^{-1} x \cdot \dfrac{1}{x \cdot \log x} + \log(\log x) \cdot \dfrac{1}{\sqrt{x^2-1}}$

$$\Rightarrow \dfrac{dy}{dx} = (\log x)^{\cosh^{-1} x} \left[\dfrac{\cosh^{-1} x}{x \log x} + \dfrac{\log(\log x)}{\sqrt{x^2-1}}\right]$$

5. $y = \dfrac{(3x^2-1)(x-5)}{(x^3+1)(2x+1)}$; Taking logarithm to the base e both sides,

$\log y = \log \dfrac{(3x^2-1)(x-5)}{(x^3+1)(2x+1)}$

$\Rightarrow \log y = \log(3x^2-1) + \log(x-5) - \log(x^3+1) - \log(2x+1)$

Differentiating w.r.to x,

$\dfrac{1}{y} \cdot \dfrac{dy}{dx} = \dfrac{6x}{3x^2-1} + \dfrac{1}{x-5} - \dfrac{3x^2}{x^3+1} - \dfrac{2}{2x+1}$

$\Rightarrow \dfrac{dy}{dx} = \dfrac{(3x^2-1)(x-5)}{(x^3+1)(2x+1)} \left[\dfrac{6x}{3x^2-1} + \dfrac{1}{x-5} - \dfrac{3x^2}{x^3+1} - \dfrac{2}{2x+1}\right]$

Correct the errors if any :

1. $y = x^{\sinh x}$; Taking logarithm to the base e both sides,

$\log y = \log x^{\sinh x} \Rightarrow \log y = \sinh x \cdot \log x$

Differentiating w.r.to x, $\dfrac{1}{y} \cdot \dfrac{dy}{dx} = \sinh x \cdot \left(\dfrac{1}{x}\right) + \log x \cdot \cosh x$

$\Rightarrow \dfrac{dy}{dx} = y\left(\dfrac{\sinh x}{x} + \log x \cdot \cosh x\right) = x^{\sinh x}\left(\dfrac{\sinh x}{x} + \log x \cdot \cosh x\right)$

2. $(\sin x)^y = (\sin y)^x$; Taking logarithm to the base e both sides,

$\log(\sin x)^y = \log(\sin y)^x \Rightarrow y \log(\sin x) = x \log(\sin y)$

Differentiating w.r.to x, $\dfrac{y \cos x}{\sin x} + \log(\sin x) \cdot \dfrac{dy}{dx} = \dfrac{x \cos y}{\sin y} + \log(\sin y)$

$\Rightarrow \log(\sin x) \cdot \dfrac{dy}{dx} = x \cdot \cot y + \log(\sin y) - y \cdot \cot x$

$$\Rightarrow \frac{dy}{dx} = \frac{x \cdot \cot y + \log(\sin y) - y \cdot \cot x}{\log(\sin x)}$$

3. $y = (x^3 - 2x + 1)^{\sqrt{x+5}}$; Taking logarithm to the base e both sides,

$$\log y = \log[(x^3 - 2x + 1)^{\sqrt{x+5}}] \Rightarrow \log y = \sqrt{x+5} \cdot \log(x^3 - 2x + 1)$$

Differentiating w.r.to x, $\quad \dfrac{1}{y} \cdot \dfrac{dy}{dx} = \dfrac{\sqrt{x+5}}{x^3 - 2x + 1} + \dfrac{\log(x^3 - 2x + 1)}{2\sqrt{x+5}}$

$$\Rightarrow \frac{dy}{dx} = (x^3 - 2x + 1)^{\sqrt{x+5}} \cdot \left[\frac{\sqrt{x+5}}{x^3 - 2x + 1} + \frac{\log(x^3 - 2x + 1)}{2\sqrt{x+5}}\right]$$

4. $y = x^{x^{x^{\cdot^{\cdot^{\cdot^\infty}}}}}$

$y = x^{x^{x^{\cdot^{\cdot^{\cdot^\infty}}}}} = x^y$; Taking logarithm to the base e both sides,

$\log y = \log x^y \Rightarrow \log y = y \cdot \log x$

Differentiating w.r.to x, $\quad \dfrac{1}{y} \cdot \dfrac{dy}{dx} = y \cdot \dfrac{1}{x} + \log x \cdot \dfrac{dy}{dx}$

$\left(\dfrac{1}{y} - \log x\right) \cdot \dfrac{dy}{dx} = \dfrac{y}{x} \Rightarrow \dfrac{dy}{dx} = \dfrac{y^2}{x(1 - y \cdot \log x)}$

5. $y = \sqrt[5]{\dfrac{2x^2 - 7}{3x - 5}}$; Taking logarithm to the base e both sides,

$\log y = \log\left(\dfrac{2x^2 - 7}{3x - 5}\right)^{1/5}$

$\Rightarrow \log y = \dfrac{1}{5}[\log(2x^2 - 7) - \log(3x - 5)]$

Differentiating w.r.to x,

$\dfrac{1}{y} \cdot \dfrac{dy}{dx} = \dfrac{1}{5}\left[\dfrac{4x}{2x^2 - 7} - \dfrac{3}{3x - 5}\right] = \dfrac{12x^2 - 20x - 6x^2 + 21}{5(2x^2 - 7)(3x - 5)} = \dfrac{6x^2 - 20x + 21}{5(2x^2 - 7)(3x - 5)}$

$\Rightarrow \dfrac{dy}{dx} = \sqrt[5]{\dfrac{2x^2 - 7}{3x - 5}} \cdot \dfrac{6x^2 - 20x + 21}{5(2x^2 - 7)(3x - 5)}$

CALCULUS I – Differential Calculus Formulae Practice Workbook Vol II

Corrected Solutions

2. CS: $\log(\sin x)^y = \log(\sin y)^x$

$\Rightarrow y \log(\sin x) = x \log(\sin y)$

Differentiating w.r.to x, $\dfrac{y \cos x}{\sin x} + \log(\sin x) \cdot \dfrac{dy}{dx} = \dfrac{x \cos y}{\sin y} \cdot \dfrac{dy}{dx} + \log(\sin y)$

$\Rightarrow \dfrac{dy}{dx}[\log(\sin x) - x \cdot \cot y] = \log(\sin y) - y \cdot \cot x$

$\Rightarrow \dfrac{dy}{dx} = \dfrac{\log(\sin y) - y \cdot \cot x}{\log(\sin x) - x \cdot \cot y}$

3. CS: $\log y = \log[(x^3 - 2x + 1)^{\sqrt{x+5}}]$

$\Rightarrow \log y = \sqrt{x+5} \cdot \log(x^3 - 2x + 1)$

Differentiating w.r.to x, $\dfrac{1}{y} \cdot \dfrac{dy}{dx} = \dfrac{\sqrt{x+5} \cdot (3x^2 - 2)}{x^3 - 2x + 1} + \dfrac{\log(x^3 - 2x + 1)}{2\sqrt{x+5}}$

$\Rightarrow \dfrac{dy}{dx} = (x^3 - 2x + 1)^{\sqrt{x+5}} \cdot \left[\dfrac{(3x^2 - 2) \cdot \sqrt{x+5}}{x^3 - 2x + 1} + \dfrac{\log(x^3 - 2x + 1)}{2\sqrt{x+5}}\right]$

1,4,5 – No error

Do it yourself :

1. $(\cot x)^{\cosh x}$ 2. $(\cos x)^y = (\cos y)^x$ 3. $(\tan^{-1} x)^{\log x}$ 4. $xy = e^{xy}$

5. $\sqrt{\dfrac{(x-1)(x+2)}{(x-3)(x+4)(x-5)}}$

Answers :

1. $(\cot x)^{\cosh x}\left[\log(\cot x) \cdot \sinh x - \dfrac{\cosh x}{\sin x \cdot \cos x}\right]$

2. $\dfrac{y \cdot \tan x + \log(\cos y)}{x \cdot \tan y + \log(\cos x)}$

3. $(\tan^{-1} x)^{\log x}\left[\dfrac{\log x}{(1+x^2)\tan^{-1} x} + \dfrac{\log(\tan^{-1} x)}{x}\right]$ 4. $\dfrac{-y}{x}$

CALCULUS I – Differential Calculus Formulae Practice Workbook Vol II

5. $\dfrac{1}{2}\sqrt{\dfrac{(x-1)(x+2)}{(x-3)(x+4)(x-5)}} \cdot [\dfrac{1}{x-1} + \dfrac{1}{x+2} - \dfrac{1}{x-3} - \dfrac{1}{x+4} - \dfrac{1}{x-5}]$

Special Case :

Find the **error** in the following.

Problem : Differentiate $x^y + y^x$ w.r.to x

Solution: Let $y = x^y + y^x$; Taking logarithm to the base e both sides,

$\log y = \log(x^y + y^x) = \log(x^y) + \log(y^x) = y.\log x + x.\log y$

Differentiating w.r.to x, $\dfrac{1}{y} \cdot \dfrac{dy}{dx} = \dfrac{y}{x} + \log x \cdot \dfrac{dy}{dx} + \dfrac{x}{y} \cdot \dfrac{dy}{dx}$

[*Mistake is committed already! Not yet found it?*]

The formula (indeed there is no formulae as such) applied in step 2 is wrong.

ie. $\log(x^y + y^x) = \log(x^y) + \log(y^x)$ is wrong.

Similarly . $\log(x^y - y^x) = \log(x^y) - \log(y^x)$ is also wrong.

We have $\log MN = \log M + \log N$ and $\log \dfrac{M}{N} = \log M - \log N$ only.

Then how to resolve it?

Corrected Solution :

$y = x^y + y^x$; Let $u = x^y$ and $v = y^x$

Then $y = u + v \Rightarrow \dfrac{dy}{dx} = \dfrac{du}{dx} + \dfrac{dv}{dx}$ ……………❶

$u = x^y$ and $v = y^x$; By taking logarithm to the base e,

we get $\log u = \log x^y$ and $\log v = \log y^x$

$\log u = y \log x$	$\log v = x \log y$
Differentiating w.r.to x	Differentiating w.r.to x
$\dfrac{1}{u} \cdot \dfrac{du}{dx} = \dfrac{y}{x} + \log x \cdot \dfrac{dy}{dx}$	$\dfrac{1}{v} \cdot \dfrac{dv}{dx} = \dfrac{x}{y} \cdot \dfrac{dy}{dx} + \log y$
$\Rightarrow \dfrac{du}{dx} = u\left[\dfrac{y}{x} + \log x \cdot \dfrac{dy}{dx}\right]$	$\Rightarrow \dfrac{dv}{dx} = v\left[\dfrac{x}{y} \cdot \dfrac{dy}{dx} + \log y\right]$
$\Rightarrow \dfrac{du}{dx} = x^y\left[\dfrac{y}{x} + \log x \cdot \dfrac{dy}{dx}\right]$	$\Rightarrow \dfrac{dv}{dx} = y^x\left[\dfrac{x}{y} \cdot \dfrac{dy}{dx} + \log y\right]$

❶ becomes $\dfrac{dy}{dx} = x^y\left[\dfrac{y}{x} + \log x \cdot \dfrac{dy}{dx}\right] + y^x\left[\dfrac{x}{y} \cdot \dfrac{dy}{dx} + \log y\right]$

$\Rightarrow (1 - x^y \log x - x \cdot y^{x-1})\dfrac{dy}{dx} = y \cdot x^{y-1} + y^x \log y$

$\Rightarrow \dfrac{dy}{dx} = \dfrac{y \cdot x^{y-1} + y^x \log y}{1 - x^y \log x - x \cdot y^{x-1}}$

2. Differentiate $(\sin x)^x + x^{\cos x}$ w.r.to x

Let $y = (\sin x)^x - x^{\cos x}$, $u = (\sin x)^x$ and $v = x^{\cos x}$

Then $y = u + v \Rightarrow \dfrac{dy}{dx} = \dfrac{du}{dx} + \dfrac{dv}{dx}$ ....❶

$u = (\sin x)^x$ and $v = x^{\cos x} \Rightarrow \log u = x \cdot \log(\sin x)$; $\log v = \cos x \cdot \log x$

$\log u = x \cdot \log(\sin x)$; Differentiating w.r.to x
$\dfrac{1}{u} \cdot \dfrac{du}{dx} = x \cdot \cot x + \log(\sin x) \Rightarrow \dfrac{du}{dx} = u\left[x \cdot \cot x + \log(\sin x)\right]$
$\Rightarrow \dfrac{du}{dx} = (\sin x)^x [x \cdot \cot x + \log(\sin x)]$
$\log v = \cos x \cdot \log x$; Differentiating w.r.to x
$\dfrac{1}{v} \cdot \dfrac{dv}{dx} = \dfrac{\cos x}{x} - \sin x \cdot \log x \Rightarrow \dfrac{dv}{dx} = v\left[\dfrac{\cos x}{x} - \sin x \cdot \log x\right]$
$\Rightarrow \dfrac{dv}{dx} = x^{\cos x}\left[\dfrac{\cos x}{x} - \sin x \cdot \log x\right]$

CALCULUS I – Differential Calculus Formulae Practice Workbook Vol II

❶ becomes

$$\frac{dy}{dx} = (\sin x)^x [x.\cot x + \log(\sin x)] + x^{\cos x} \left[\frac{\cos x}{x} - \sin x . \log x\right]$$

Do it yourself : Find the derivative of the following:

1. $x^{\cos x} + \frac{x^2+1}{x^2-1}$

2. $(\sec x)^x + (x \sin x)^{\frac{1}{x}}$

3. What are the different methods known to you to find the derivative of $(x+2)^2(x^2 - 3x + 1)$ w.r.to x. Find it's deivative.

Answers :

1. $x^{\cos x}\left[\frac{\cos x}{x} - \sin x . \log x\right] - \frac{4x}{(x^2-1)^2}$

2. $(\sec x)^x [x \tan x + \log \sec x] + \frac{(x \sin x)^{\frac{1}{x}}}{x^2} [x \cot x + 1 - \log(x \sin x)]$

3. by a) reducing into a single polynomial in x

   b) using the product rule.

   c) using logarithmic method.

   It's derivative w.r.to x is $4x^3 + 3x^2 - 14x - 8$

Unit 9 - Derivative of a function w.r.to another function

Examples :

1. Differentiate e^x w.r.to x^2

Let $u = e^x$ and $v = x^2$. It is enough to find $\dfrac{du}{dv}$ ie. $\dfrac{du}{dx} \div \dfrac{dv}{dx}$

Here $\dfrac{du}{dx} = e^x$; $\dfrac{dv}{dx} = 2x$.

Hence $\dfrac{du}{dv} = \dfrac{e^x}{2x}$

2. Differentiate $\tan^{-1}\dfrac{x}{\sqrt{1-x^2}}$ w.r.to $\sec^{-1}\dfrac{1}{2x^2-1}$

Let $u = \tan^{-1}\dfrac{x}{\sqrt{1-x^2}}$ and $v = \sec^{-1}\dfrac{1}{2x^2-1}$. It is enough to find $\dfrac{du}{dv}$.

Substituting $x = \sin\theta$, we get $u = \tan^{-1}\dfrac{\sin\theta}{\cos\theta} = \tan^{-1}(\tan\theta) = \theta$ and

$v = \sec^{-1}\dfrac{1}{2\sin^2\theta - 1} = \sec^{-1}(-\sec 2\theta)$

$= \sec^{-1}[\sec(180 - 2\theta)] = 180 - 2\theta$

Hence, $u = \theta$ and $v = 180 - 2\theta$; Differentiating u and v w.r.to θ,

we get $\dfrac{du}{d\theta} = 1$; $\dfrac{dv}{d\theta} = -2$ and $\dfrac{du}{dv} = -\dfrac{1}{2}$

Exercise :

1. Differentiate $\sin 2x$ w.r.to $\sqrt{x}$

2. Differentiate $\tan^{-1}\dfrac{\sqrt{1+x^2}-1}{x}$ w.r.to $\tan^{-1} x$

Solution :

1. Differentiate $\sin 2x$ w.r.to $\sqrt{x}$

Let $u = \sin 2x$ and $v = \sqrt{x}$. It is enough to find $\dfrac{du}{dv}$ ie. $\dfrac{du}{dx} \div \dfrac{dv}{dx}$

Here $\dfrac{du}{dx} = 2\cos 2x$; $\dfrac{dv}{dx} = \dfrac{1}{2\sqrt{x}}$.

Hence $\dfrac{du}{dv} = 4\sqrt{x}\cos 2x$

2. Differentiate $\tan^{-1}\dfrac{\sqrt{1+x^2}-1}{x}$ w.r.to $\tan^{-1} x$

Let $u = \tan^{-1}\dfrac{\sqrt{1+x^2}-1}{x}$ and $v = \tan^{-1} x$. It is enough to find $\dfrac{du}{dv}$.

Substituting $x = \tan\theta$, we get,

$$u = \tan^{-1}\left(\dfrac{\sqrt{1+\tan^2\theta}-1}{\tan\theta}\right) = \tan^{-1}\left(\dfrac{\sec\theta - 1}{\tan\theta}\right) = \tan^{-1}\left(\dfrac{1-\cos\theta}{\sin\theta}\right)$$

$$= \tan^{-1}\left(\dfrac{2\sin^2\frac{\theta}{2}}{2\sin\frac{\theta}{2}\cos\frac{\theta}{2}}\right) = \tan^{-1}\left(\tan\dfrac{\theta}{2}\right) = \dfrac{\theta}{2}$$

and $v = \tan^{-1}(\tan\theta) = \theta$

Hence, $u = \dfrac{\theta}{2}$ and $v = \theta$; Differentiating u and v w.r.to θ,

we get $\dfrac{du}{d\theta} = \dfrac{1}{2}$; $\dfrac{dv}{d\theta} = 1$ and $\dfrac{du}{dv} = \dfrac{1}{2}$

Find the errors if any:

1. Differentiate $a^{\sin x}$ w.r.to $\cos x$

Let $u = a^{\sin x}$ and $v = \cos x$. It is enough to find $\dfrac{du}{dv}$ ie. $\dfrac{du}{dx} \div \dfrac{dv}{dx}$

Here $\dfrac{du}{dx} = a^{\sin x} \cdot \log_a e \cdot \cos x$; $\dfrac{dv}{dx} = -\sin x$.

Hence $\dfrac{du}{dv} = -\cot x \cdot a^{\sin x} \cdot \log_a e$

2. Differentiate $\log_{10} x$ w.r.to x^2

Let $u = \log_{10} x$ and $v = x^2$. It is enough to find $\frac{du}{dv}$ ie. $\frac{du}{dx} \div \frac{dv}{dx}$

Here $\frac{du}{dx} = \frac{1}{x} \log_{10} e$ $\frac{dv}{dx} = 2x$.

Hence $\frac{du}{dv} = \frac{1}{2x^2} \log_{10} e$

Corrected Solution

1. **C.S** : $\frac{du}{dx} = a^{\sin x} . \log_e a . \cos x ;$ $\frac{dv}{dx} = -\sin x$.

Hence $\frac{du}{dv} = -\cot x . a^{\sin x} . \log_e a$ 2. No error

Do it yourself:

1. Differentiate $e^{\tan^{-1} x}$ w.r.to $\tan^{-1} x$

2. Differentiate $\tan^{-1} \frac{2x}{1-x^2}$ w.r.to $\sin^{-1} \frac{2x}{1+x^2}$

Answers: 1. $e^{\tan^{-1} x}$ 2. 1

CALCULUS I – Differential Calculus Formulae Practice Workbook Vol II

Unit 10 Differentiation Technique :

We have so far discussed in detail the methods available in Differential Calculus. As already mentioned, we have not discussed the theory part. Apart from this, something is irking in our mind. What is that?

We have applied many formulae while differentiating various functions without asking any question. Isn't it? How did these formulae come?

Here we discuss the techniques used to derive the basic formulae. The technique used here is known as 'first principle technique'. It is based on the definition of derivative discussed already in volume I.

Definition of Derivative : If $y = f(x)$, $\dfrac{dy}{dx} = \lim_{\Delta x \to 0} \dfrac{f(x+\Delta x) - f(x)}{\Delta x}$ where Δx is a small increment in x.

10.1 Rule of 'first principle': Let $y = f(x)$ be the given function.

1. Increase the argument x by Δx. As y is a function of x, there will be a change in y also. Let the corresponding increment in y be Δy. Then $y + \Delta y = f(x + \Delta x)$

   [Note that $y + \Delta y = f(x) + \Delta x$ is wrong.]

2. From this, the value of $\Delta y = f(x + \Delta x) - y = f(x + \Delta x) - f(x)$

3. Dividing both sides by Δx, we get $\dfrac{\Delta y}{\Delta x} = \dfrac{f(x + \Delta x) - f(x)}{\Delta x}$

4. Taking $\lim \Delta x \to 0$ both sides, we get $\lim_{\Delta x \to 0} \dfrac{\Delta y}{\Delta x} = \lim_{\Delta x \to 0} \dfrac{f(x + \Delta x) - f(x)}{\Delta x}$

5. By the definition of derivative, $\dfrac{dy}{dx} = \lim_{\Delta x \to 0} \dfrac{f(x + \Delta x) - f(x)}{\Delta x}$

Note : As $y = f(x)$, the above can be written as $\dfrac{dy}{dx} = \lim_{\Delta x \to 0} \dfrac{(y + \Delta y) - y}{\Delta x}$

CALCULUS I – Differential Calculus Formulae Practice Workbook Vol II

Formulae required here:

A. Limit of a function :

1. For $\left|\frac{\Delta x}{a}\right| < 1$ and for any rational index n, $\lim_{x \to a} \frac{(x+a)^n - a^n}{x-a} = na^{n-1}$, $(a \neq 0)$

2. $\lim_{\theta \to 0} \frac{\sin \theta}{\theta} = 1$ 　　3. $\lim_{\theta \to 0} \frac{\tan \theta}{\theta} = 1$ 　　4. $\lim_{t \to 0} \frac{e^t - 1}{t} = 1$

B. Trigonometry :

1. $\sin(A - B) = \sin A \cos B - \cos A \sin B$

2. $\sin C - \sin D = 2 \cos \frac{C+D}{2} \sin \frac{C-D}{2}$ 　　3. $\cos C - \cos D = -2 \sin \frac{C+D}{2} \sin \frac{C-D}{2}$

Examples:

1. Find the derivative of x^n where n is any rational number using first principles.

Let $y = x^n$; Let Δx be a small increment in x.

Let the corresponding increment in y be Δy.

Now, $y + \Delta y = (x + \Delta x)^n \Rightarrow \Delta y = (x + \Delta x)^n - y = (x + \Delta x)^n - x^n$

$$\Rightarrow \frac{\Delta y}{\Delta x} = \frac{(x + \Delta x)^n - x^n}{\Delta x}$$

$$\Rightarrow \lim_{\Delta x \to 0} \frac{\Delta y}{\Delta x} = \lim_{\Delta x \to 0} \frac{(x + \Delta x)^n - x^n}{(x + \Delta x) - x}$$

$$\Rightarrow \frac{dy}{dx} = nx^{n-1} \text{ using the formula A.1}$$

2. Find the derivative of C where C is a constant using first principles

Let $y = C$; As C is a constant, it's increment is 0.

Now, $y + \Delta y = C \Rightarrow \Delta y = C - C = 0$

$$\Rightarrow \frac{\Delta y}{\Delta x} = 0 \Rightarrow \lim_{\Delta x \to 0} \frac{\Delta y}{\Delta x} = 0 \Rightarrow \frac{dy}{dx} = 0$$

3. Find the derivative of e^x using first principles.

Let $y = e^x$; Let Δx be a small increment in x.

CALCULUS I – Differential Calculus Formulae Practice Workbook Vol II

Let the corresponding increment in y be Δy.

Now, $y + \Delta y = e^{x+\Delta x} \implies \Delta y = e^{x+\Delta x} - y = e^{x+\Delta x} - e^x$

$$\implies \frac{\Delta y}{\Delta x} = \frac{e^x(e^{\Delta x} - 1)}{\Delta x}$$

$$\implies \lim_{\Delta x \to 0} \frac{\Delta y}{\Delta x} = e^x \lim_{\Delta x \to 0} \frac{e^{\Delta x} - 1}{\Delta x}$$

$$\implies \frac{dy}{dx} = e^x \text{ using the formula A.4}$$

4. Find the derivative of $\sin x$ using first principles.

Let $y = \sin x$; Let Δx be a small increment in x.

Let the corresponding increment in y be Δy.

Now, $y + \Delta y = \sin(x + \Delta x) \implies \Delta y = \sin(x + \Delta x) - y = \sin(x + \Delta x) - \sin x$

$$\implies \frac{\Delta y}{\Delta x} = \frac{\sin(x + \Delta x) - \sin x}{\Delta x} \implies \lim_{\Delta x \to 0} \frac{\Delta y}{\Delta x} = \lim_{\Delta x \to 0} \frac{2 \cos(x + \frac{\Delta x}{2}) \cdot \sin\frac{\Delta x}{2}}{\Delta x} \text{ using B.2}$$

$$\implies \lim_{\Delta x \to 0} \frac{\Delta y}{\Delta x} = \lim_{\Delta x \to 0} \cos(x + \frac{\Delta x}{2}) \cdot \lim_{\Delta x \to 0} \frac{\sin\frac{\Delta x}{2}}{\frac{\Delta x}{2}}$$

$$\implies \frac{dy}{dx} = \cos x \text{ using the formula A.2}$$

5. Find the derivative of $\cos x$ using first principles.

Let $y = \cos x$; Let Δx be a small increment in x.

Let the corresponding increment in y be Δy.

Now, $y + \Delta y = \cos(x + \Delta x) \implies \Delta y = \cos(x + \Delta x) - y = \cos(x + \Delta x) - \cos x$

$$\implies \frac{\Delta y}{\Delta x} = \frac{\cos(x + \Delta x) - \cos x}{\Delta x} \implies \lim_{\Delta x \to 0} \frac{\Delta y}{\Delta x} = \lim_{\Delta x \to 0} \frac{-2 \sin(x + \frac{\Delta x}{2}) \cdot \sin\frac{\Delta x}{2}}{\Delta x} \text{ using B.3}$$

$$\implies \lim_{\Delta x \to 0} \frac{\Delta y}{\Delta x} = - \lim_{\Delta x \to 0} \sin(x + \frac{\Delta x}{2}) \cdot \lim_{\Delta x \to 0} \frac{\sin\frac{\Delta x}{2}}{\frac{\Delta x}{2}}$$

$$\implies \frac{dy}{dx} = - \sin x \text{ using the formula A.2}$$

CALCULUS I – Differential Calculus Formulae Practice Workbook Vol II

6. Find the derivative of $\tan x$ using first principles.

Let $y = \tan x = \dfrac{\sin x}{\cos x}$; Let Δx be a small increment in x.

Let the corresponding increment in y be Δy.

Now, $y + \Delta y = \dfrac{\sin(x+\Delta x)}{\cos(x+\Delta x)} \implies \Delta y = \dfrac{\sin(x+\Delta x)}{\cos(x+\Delta x)} - \dfrac{\sin x}{\cos x}$

$$\implies \dfrac{\Delta y}{\Delta x} = \dfrac{\sin(x+\Delta x).\cos x - \cos(x+\Delta x).\sin x}{\cos(x+\Delta x).\cos x.\,\Delta x}$$

$$\implies \lim_{\Delta x \to 0}\dfrac{\Delta y}{\Delta x} = \lim_{\Delta x \to 0}\dfrac{1}{\cos(x+\Delta x).\cos x} \cdot \lim_{\Delta x \to 0}\dfrac{\sin \Delta x}{\Delta x}$$

$$\implies \dfrac{dy}{dx} = \dfrac{1}{\cos^2 x} = \sec^2 x$$

7. Find the derivative of $\operatorname{cosec} x$ using first principles.

Let $y = \operatorname{cosec} x = \dfrac{1}{\sin x}$; Let Δx be a small increment in x.

Let the corresponding increment in y be Δy.

Now, $y + \Delta y = \dfrac{1}{\sin(x+\Delta x)} \implies \Delta y = \dfrac{1}{\sin(x+\Delta x)} - \dfrac{1}{\sin x}$

$$\implies \dfrac{\Delta y}{\Delta x} = \dfrac{\sin x - \sin(x+\Delta x)}{\sin(x+\Delta x).\sin x.\,\Delta x}$$

$$\implies \dfrac{\Delta y}{\Delta x} = \dfrac{2\cos\left(x+\frac{\Delta x}{2}\right).\sin\left(-\frac{\Delta x}{2}\right)}{\sin(x+\Delta x).\sin x.\,\Delta x}$$

$$\implies \lim_{\Delta x \to 0}\dfrac{\Delta y}{\Delta x} = \lim_{\Delta x \to 0}\dfrac{-\cos\left(x+\frac{\Delta x}{2}\right)}{\sin(x+\Delta x).\sin x} \cdot \lim_{\Delta x \to 0}\dfrac{\sin\frac{\Delta x}{2}}{\frac{\Delta x}{2}}$$

$$\implies \dfrac{dy}{dx} = \dfrac{-\cos x}{\sin^2 x} = -\operatorname{cosec} x . \cot x$$

8. Find the derivative of the $y = u.v$ where u and v are differentiable functions of x.

$y = u(x).v(x) \implies y + \Delta y = u(x+\Delta x).v(x+\Delta x)$

$$\implies \Delta y = u(x+\Delta x).v(x+\Delta x) - u(x).v(x)$$

CALCULUS I – Differential Calculus Formulae Practice Workbook Vol II

$$\Rightarrow \frac{\Delta y}{\Delta x} = \frac{u(x + \Delta x) \cdot v(x + \Delta x) - u(x) \cdot v(x)}{\Delta x}$$

Adding and subtracting $u(x + \Delta x) \cdot v(x)$ in the numerator of the R.H.S we get

$$\Rightarrow \frac{\Delta y}{\Delta x} = \frac{u(x + \Delta x) \cdot v(x + \Delta x) + u(x + \Delta x) \cdot v(x) - u(x + \Delta x) \cdot v(x) - u(x) \cdot v(x)}{\Delta x}$$

$$\Rightarrow \frac{\Delta y}{\Delta x} = \frac{u(x + \Delta x)[v(x + \Delta x) - v(x)] + v(x)[u(x + \Delta x) - u(x)]}{\Delta x}$$

$$\Rightarrow \lim_{\Delta x \to 0} \frac{\Delta y}{\Delta x} = \lim_{\Delta x \to 0} u(x + \Delta x) \cdot \lim_{\Delta x \to 0} \frac{v(x + \Delta x) - v(x)}{\Delta x}$$

$$+ v(x) \lim_{\Delta x \to 0} \frac{u(x + \Delta x) - u(x)}{\Delta x}$$

$$\Rightarrow \frac{dy}{dx} = u(x) \cdot \frac{d[v(x)]}{dx} + v(x) \cdot \frac{d[u(x)]}{dx} = u(x) \cdot v'(x) + v(x) \cdot u'(x) = u \cdot v' + v \cdot u'$$

Exercise : Find the derivative of the following w.r.to x using first principles.

1. x^{15} 2. $\sqrt{x}$ 3. e^{3x} 4. $\sin 2x$ 5. $\cos 5x$

6. $\cot x$ 7. $\sec x$

Solutions :

1. Let $y = x^{15}$; Let Δx be a small increment in x.

   Let the corresponding increment in y be Δy.

   Now, $y + \Delta y = (x + \Delta x)^{15} \Rightarrow \Delta y = (x + \Delta x)^{15} - y = (x + \Delta x)^{15} - x^{15}$

   $$\Rightarrow \frac{\Delta y}{\Delta x} = \frac{(x + \Delta x)^{15} - x^{15}}{\Delta x}$$

   $$\Rightarrow \lim_{\Delta x \to 0} \frac{\Delta y}{\Delta x} = \lim_{\Delta x \to 0} \frac{(x + \Delta x)^{15} - x^{15}}{(x + \Delta x) - x}$$

   $$\Rightarrow \frac{dy}{dx} = 15 x^{14} \quad \text{using the formula A.1}$$

2. Let $y = x^{\frac{1}{2}}$; Let Δx be a small increment in x.

   Let the corresponding increment in y be Δy.

   Now, $y + \Delta y = (x + \Delta x)^{\frac{1}{2}} \Rightarrow \Delta y = (x + \Delta x)^{\frac{1}{2}} - y = (x + \Delta x)^{\frac{1}{2}} - x^{\frac{1}{2}}$

$$\Rightarrow \frac{\Delta y}{\Delta x} = \frac{(x+\Delta x)^{\frac{1}{2}} - x^{\frac{1}{2}}}{\Delta x}$$

$$\Rightarrow \lim_{\Delta x \to 0} \frac{\Delta y}{\Delta x} = \lim_{\Delta x \to 0} \frac{(x+\Delta x)^{\frac{1}{2}} - x^{\frac{1}{2}}}{(x+\Delta x) - x}$$

$$\Rightarrow \frac{dy}{dx} = \frac{1}{2} x^{-\frac{1}{2}} = \frac{1}{2\sqrt{x}} \text{ using the formula A.1}$$

3. Let $y = e^{3x}$; Let Δx be a small increment in x.

Let the corresponding increment in y be Δy.

Now, $y + \Delta y = e^{3(x+\Delta x)} \Rightarrow \Delta y = e^{3(x+\Delta x)} - y = e^{3(x+\Delta x)} - e^{3x}$

$$\Rightarrow \frac{\Delta y}{\Delta x} = \frac{e^{3x}(e^{3\Delta x} - 1)}{\Delta x}$$

$$\Rightarrow \lim_{\Delta x \to 0} \frac{\Delta y}{\Delta x} = 3 e^{3x} \lim_{\Delta x \to 0} \frac{e^{3\Delta x} - 1}{3 \Delta x}$$

$$\Rightarrow \frac{dy}{dx} = 3 e^{3x} \text{ using the formula A.4}$$

4. Let $y = \sin 2x$; Let Δx be a small increment in x.

Let the corresponding increment in y be Δy.

$y + \Delta y = \sin 2(x + \Delta x) \Rightarrow \Delta y = \sin 2(x + \Delta x) - y = \sin 2(x + \Delta x) - \sin 2x$

$$\Rightarrow \frac{\Delta y}{\Delta x} = \frac{\sin 2(x + \Delta x) - \sin 2x}{\Delta x} \Rightarrow \lim_{\Delta x \to 0} \frac{\Delta y}{\Delta x} = \lim_{\Delta x \to 0} \frac{2 \cos(2x + \Delta x) \cdot \sin \Delta x}{\Delta x}$$

$$\Rightarrow \lim_{\Delta x \to 0} \frac{\Delta y}{\Delta x} = 2 \lim_{\Delta x \to 0} \cos(2x + \Delta x) \cdot \lim_{\Delta x \to 0} \frac{\sin \Delta x}{\Delta x}$$

$$\Rightarrow \frac{dy}{dx} = 2 \cos 2x \text{ using the formula A.2}$$

5. Let $y = \cos 5x$; Let Δx be a small increment in x.

Let the corresponding increment in y be Δy.

$y + \Delta y = \cos 5(x + \Delta x) \Rightarrow \Delta y = \cos 5(x + \Delta x) - \cos 5x$

$$\Rightarrow \frac{\Delta y}{\Delta x} = \frac{\cos 5(x + \Delta x) - \cos 5x}{\Delta x}$$

$$\Rightarrow \lim_{\Delta x \to 0} \frac{\Delta y}{\Delta x} = \lim_{\Delta x \to 0} \frac{-2\sin(5x + \frac{5\Delta x}{2}) \cdot \sin\frac{5\Delta x}{2}}{\Delta x} \text{ using B.2}$$

$$\Rightarrow \lim_{\Delta x \to 0} \frac{\Delta y}{\Delta x} = -5 \lim_{\Delta x \to 0} \sin(5x + \frac{5\Delta x}{2}) \cdot \lim_{\Delta x \to 0} \frac{\sin\frac{5\Delta x}{2}}{\frac{5\Delta x}{2}}$$

$$\Rightarrow \frac{dy}{dx} = -5\sin 5x$$

6. Let $y = \cot x = \dfrac{\cos x}{\sin x}$; Let Δx be a small increment in x.

Let the corresponding increment in y be Δy.

Now, $y + \Delta y = \dfrac{\cos(x+\Delta x)}{\sin(x+\Delta x)} \Rightarrow \Delta y = \dfrac{\cos(x+\Delta x)}{\sin(x+\Delta x)} - \dfrac{\cos x}{\sin x}$

$$\Rightarrow \frac{\Delta y}{\Delta x} = \frac{\sin x \cdot \cos(x + \Delta x) - \cos x \cdot \sin(x+\Delta x)}{\sin(x+\Delta x) \cdot \sin x \cdot \Delta x}$$

$$\Rightarrow \frac{\Delta y}{\Delta x} = \frac{-\sin \Delta x}{\sin(x+\Delta x) \cdot \sin x \cdot \Delta x}$$

$$\Rightarrow \lim_{\Delta x \to 0} \frac{\Delta y}{\Delta x} = \lim_{\Delta x \to 0} \frac{-1}{\sin(x+\Delta x) \cdot \sin x} \cdot \lim_{\Delta x \to 0} \frac{\sin \Delta x}{\Delta x}$$

$$\Rightarrow \frac{dy}{dx} = \frac{-1}{\sin^2 x} = -\csc^2 x$$

7. Find the derivative of $\sec x$ using first principles.

Let $y = \sec x = \dfrac{1}{\cos x}$; Let Δx be a small increment in x.

Let the corresponding increment in y be Δy.

Now, $y + \Delta y = \dfrac{1}{\cos(x+\Delta x)} \Rightarrow \Delta y = \dfrac{1}{\cos(x + \Delta x)} - \dfrac{1}{\cos x}$

$$\Rightarrow \frac{\Delta y}{\Delta x} = \frac{\cos x - \cos(x + \Delta x)}{\cos(x+\Delta x) \cdot \cos x \cdot \Delta x} \Rightarrow \frac{\Delta y}{\Delta x} = \frac{-2\sin\left(x + \frac{\Delta x}{2}\right) \cdot \sin\left(-\frac{\Delta x}{2}\right)}{\cos(x+\Delta x) \cdot \cos x \cdot \Delta x}$$

$$\Rightarrow \lim_{\Delta x \to 0} \frac{\Delta y}{\Delta x} = \lim_{\Delta x \to 0} \frac{\sin\left(x + \frac{\Delta x}{2}\right)}{\cos(x+\Delta x) \cdot \cos x} \cdot \lim_{\Delta x \to 0} \frac{\sin\frac{\Delta x}{2}}{\frac{\Delta x}{2}}$$

$$\Rightarrow \frac{dy}{dx} = \frac{\sin x}{\cos^2 x} = \sec x . \tan x$$

Do it yourself:

1. Find the derivative of $\sqrt[3]{x}$ using first principles.

2. Find the derivative of $\frac{1}{x}$ using first principles.

10.2 Derivative of inverse trigonometric functions:

1. $y = \sin^{-1} x$ then $x = \sin y$. Differentiating w.r.to y.

we get, $\frac{dx}{dy} = \cos y . (1) = \sqrt{1 - \sin^2 y} = \sqrt{1 - x^2} \Rightarrow \frac{dy}{dx} = \frac{1}{\sqrt{1-x^2}}$

2. $y = \tan^{-1} x$ then $x = \tan y$. Differentiating w.r.to y.

we get, $\frac{dx}{dy} = \sec^2 y = 1 + \tan^2 y = 1 + x^2 \Rightarrow \frac{dy}{dx} = \frac{1}{1 + x^2}$

3. $y = \operatorname{cosec}^{-1} x$ then $x = \operatorname{cosec} y$. Differentiating w.r.to y.

we get, $\frac{dx}{dy} = - \operatorname{cosec} y . \cot y = - \operatorname{cosec} y \sqrt{\operatorname{cosec}^2 y - 1}$

$= - x\sqrt{x^2 - 1} \Rightarrow \frac{dy}{dx} = - \frac{1}{x\sqrt{x^2 - 1}}$

Do it yourself: Find the derivative of the following w.r.to x:

1. $\cos^{-1} x$
2. $\cot^{-1} x$
3. $\sec^{-1} x$

10.3 Derivative of hyperbolic functions:

Required formulae:

1. $\sinh x = \dfrac{e^x - e^{-x}}{2}$
2. $\cosh x = \dfrac{e^x + e^{-x}}{2}$
3. $\tanh x = \dfrac{e^x - e^{-x}}{e^x + e^{-x}}$

4. $\coth x = \dfrac{e^x + e^{-x}}{e^x - e^{-x}}$
5. $\operatorname{sech} x = \dfrac{2}{e^x + e^{-x}}$
6. $\operatorname{cosech} x = \dfrac{2}{e^x - e^{-x}}$

7. $\cosh^2 x - \sinh^2 x = 1$

CALCULUS I – Differential Calculus Formulae Practice Workbook Vol II

Find the derivative of the following w.r.to x :

1. $y = \sinh x = \dfrac{e^x - e^{-x}}{2}$; Diff. w.r.to x, we get, $\dfrac{dy}{dx} = \dfrac{e^x + e^{-x}}{2} = \cosh x$

2. $y = \cosh x = \dfrac{e^x + e^{-x}}{2}$; Diff. w.r.to x, we get, $\dfrac{dy}{dx} = \dfrac{e^x - e^{-x}}{2} = \sinh x$

3. $y = \tanh x = \dfrac{\sinh x}{\cosh x}$ => $\dfrac{dy}{dx} = \dfrac{\cosh^2 x - \sinh^2 x}{\cosh^2 x} = \dfrac{1}{\cosh^2 x} = \operatorname{sech}^2 x$

4. $y = \operatorname{sech} x = \dfrac{1}{\cosh x}$ => $\dfrac{dy}{dx} = \dfrac{-\sinh x}{\cosh^2 x} = -\operatorname{sech} x . \tanh x$

Do it yourself: Find the derivative of the following w.r.to x:

1. $\coth x$ 2. $\operatorname{cosech} x$

10.4 Derivative of inverse hyperbolic functions:

Required formulae:

1. $\cosh^2 x - \sinh^2 x = 1$ 2. $\tanh^2 x + \operatorname{sech}^2 x = 1$ 3. $\coth^2 x - \operatorname{cosech}^2 x = 1$

Find the derivative of the following w.r.to x :

1. $y = \sinh^{-1} x$ then $x = \sinh y$. Differentiating w.r.to **y**.

we get, $\dfrac{dx}{dy} = \cosh y = \sqrt{1 + \sinh^2 y} = \sqrt{1 + x^2}$ => $\dfrac{dy}{dx} = \dfrac{1}{\sqrt{1+x^2}}$

2. For $|x| < 1$, $y = \tanh^{-1} x$ then $x = \tanh y$. Differentiating w.r.to **y**.

we get, $\dfrac{dx}{dy} = \operatorname{sech}^2 y = 1 - \tanh^2 y = 1 - x^2$ => $\dfrac{dy}{dx} = \dfrac{1}{1 - x^2}$

3. For $x \neq 0$, $y = \operatorname{cosech}^{-1} x$ then $x = \operatorname{cosech} y$. Differentiating w.r.to **y**.

we get, $\dfrac{dx}{dy} = -\operatorname{cosech} y . \coth y = -\operatorname{cosech} y \sqrt{\operatorname{cosech}^2 y + 1}$

$= -x\sqrt{x^2 + 1}$ => $\dfrac{dy}{dx} = -\dfrac{1}{x\sqrt{x^2+1}}$

Do it yourself: Find the derivative of the following w.r.to x:

1. $\cosh^{-1} x$ $(x > 1)$ 2. $\coth^{-1} x$ $(|x| > 1)$ 3. $\operatorname{sech}^{-1} x$ $(0 < x < 1)$

CALCULUS I – Differential Calculus Formulae Practice Workbook Vol II

Self Evaluation Test

I. Choose the correct answer

$y = f(x)$; Δx be a small increment in x and Δy be the corresponding increment in y.

1. If $y = x^7$, then $y + \Delta y = \ldots\ldots\ldots$

  a) $x^7 + \Delta x$ b) $x + \Delta x^7$ c) $(x + \Delta x)^7$ d) none of these

2. If $y = \sin 3x$, then $y + \Delta y = \ldots\ldots\ldots$

  a) $\sin 3x + \Delta x$ b) $\sin(3x + \Delta x)$ c) $\sin 3(x + \Delta x)$ d) none of these

3. If $y = \dfrac{1}{x}$, then $y + \Delta y = \ldots\ldots\ldots$

  a) $\dfrac{1}{x + \Delta x}$ b) $\dfrac{1}{x} + \Delta x$ c) $\dfrac{1}{x \, \Delta x}$ d) none of these

4. If $y = e^{2x}$, then $y + \Delta y = \ldots\ldots\ldots$

  a) $e^{2x} + \Delta x$ b) $e^{2x + \Delta x}$ c) $e^{2(x + \Delta x)}$ d) none of these

5. If $y = \sqrt[5]{x}$, then $y + \Delta y = \ldots\ldots\ldots$

  a) $\sqrt[5]{x + \Delta x}$ b) $\sqrt[5]{x} + \Delta x$ c) $\sqrt[5]{x} + \sqrt{\Delta x}$ d) none of these

6. If $y = \cos^5 2x$, then $y + \Delta y = \ldots\ldots\ldots$

  a) $\cos^5(2x + \Delta x)$ b) $\cos^5 2x + \Delta x$ c) $\cos^5 2(x + \Delta x)$ d) none of these

7. If $y = \tan^{-1} x$, then $y + \Delta y = \ldots\ldots\ldots$

  a) $\tan^{-1} x + \Delta x$ b) $\tan^{-1}(x + \Delta x)$ c) $\tan(x^{-1} + \Delta x)$ d) none of these

8. If $f(x) = \dfrac{u(x)}{v(x)}$, then $y + \Delta y = \ldots\ldots\ldots$

  a) $\dfrac{u(x)}{v(x)} + \Delta x$ b) $\dfrac{u(x) + \Delta x}{v(x)}$ c) $\dfrac{u(x + \Delta x)}{v(x + \Delta x)}$ d) none of these

9. If $y = 9$, then $y + \Delta y = \ldots\ldots\ldots$

  a) 9   b) 0   c) $9 + \Delta x$ d) none of these

10. If $y = ku(x)$, then $y + \Delta y = \ldots\ldots\ldots$ (k is a constant)

a) $ku(x) + \Delta x$ b) $k[u(x) + \Delta x]$ c) $ku(x + \Delta x)$ d) none of these

II. Choose the correct answer:

1. If $y = f(x)$, $\dfrac{dy}{dx} = \lim_{\Delta x \to 0}$ ... ...

a) $\dfrac{f(x+\Delta x)+f(x)}{\Delta x}$ b) $\dfrac{f(x+\Delta x)-f(x)}{x}$ c) $\dfrac{f(x+\Delta x)-f(x)}{\Delta x}$ d) none of these

2. $\lim_{\theta \to 0} \dfrac{\text{......}}{\theta} = 1$

a) $\sin \theta$ b) $\cos \theta$ c) $\tan \theta$ d) none of these

3. $\lim_{t \to 0}$ .... $= 1$

a) $\dfrac{e^t - t}{t}$ b) $\dfrac{e^t - 1}{t}$ c) $\dfrac{e^t - 1}{e^t}$ d) none of these

4. $2 \cos \dfrac{C+D}{2} \sin \dfrac{C-D}{2} = $ ....

a) $\cos C + \cos D$ b) $\cos C - \cos D$ c) $\sin C + \sin D$ d) $\sin C - \sin D$

5. $2 \sin \dfrac{C+D}{2} \sin \dfrac{C-D}{2} = $ ....

a) $\cos C + \cos D$ b) $\cos C - \cos D$ c) $\sin C + \sin D$ d) none of these

III. Fill in the blanks

1. If $y = \dfrac{\sin x}{1+\tan x}$, find $\dfrac{dy}{dx}$

$\dfrac{dy}{dx} = \dfrac{(1+\tan x).\cos x - \sin x.\ \text{............}}{\text{............}}$

2. If $y = \log_a \tan 2x$, find $\dfrac{dy}{dx}$

$\dfrac{dy}{dx} = \dfrac{1}{\tan 2x} \cdot 2 \sec^2 2x.$ ... ... $= $ ....

3. If $y = \sin^{-1} \dfrac{x}{\sqrt{1+x^2}}$, find $\dfrac{dy}{dx}$

Substitute $x = $........ => $\theta = $ ....

then $y = sin^{-1} \ldots\ldots = sin^{-1} sin\theta = \theta = \ldots\ldots$

Differentiating w. r. to x, $\dfrac{dy}{dx} = \dfrac{1}{1+x^2}$

4. If $x = a\cosh\theta$, $y = b\sinh\theta$, find $\dfrac{dy}{dx}$

$x = a\cosh\theta \Rightarrow \dfrac{dx}{d\theta} = \ldots\ldots$ and $y = b\sinh\theta \Rightarrow \dfrac{dy}{d\theta} = \ldots\ldots$

$\dfrac{dy}{dx} = \dfrac{\ldots\ldots}{a}$

5. If $y = sin\dfrac{x}{2} \cdot sin\dfrac{x^2}{2^2} \cdot sin\dfrac{x^3}{2^3} \cdot sin\dfrac{x^4}{2^4}$ find $\dfrac{dy}{dx}$

$y = sin\dfrac{x}{2} \cdot sin\dfrac{x^2}{2^2} \cdot sin\dfrac{x^3}{2^3} \cdot sin\dfrac{x^4}{2^4}$; Taking log both sides we get

$\log y = \ldots\ldots\ldots\ldots\ldots$

Diff. w. r. to x, $\ldots\ldots = \dfrac{\cos\dfrac{x}{2}}{2\sin\dfrac{x}{2}} + \dfrac{\ldots\ldots}{2^2\sin\dfrac{x^2}{2^2}} + \dfrac{3x^2\cos\dfrac{x^3}{2^3}}{\ldots\ldots} + \dfrac{4x^3\cos\dfrac{x^4}{2^4}}{2^4\sin\dfrac{x^4}{2^4}}$

$\dfrac{dy}{dx} = y\left[\dfrac{1}{2}\cot\dfrac{x}{2} + \dfrac{x}{2}\cot\dfrac{x^2}{2^2} + \dfrac{3x^2}{2^3}\cot\dfrac{x^3}{2^3} + \dfrac{x^3}{4}\cot\dfrac{x^4}{2^4}\right]$

where $y = sin\dfrac{x}{2} \cdot sin\dfrac{x^2}{2^2} \cdot sin\dfrac{x^3}{2^3} \cdot sin\dfrac{x^4}{2^4}$

IV. Correct the errors if any

1. $y = \tan(\sqrt{e^{\sin x}}) + \sin(e^{\sqrt{\tan x}})$

$\dfrac{dy}{dx} = \sec^2(\sqrt{e^{\sin x}}) \cdot \dfrac{1}{2\sqrt{e^{\sin x}}} \cdot e^{\sin x} \cdot \cos x$

$\qquad + \cos(e^{\sqrt{\tan x}}) \cdot e^{\sqrt{\tan x}} \cdot e^{\tan x} \cdot \sec^2 x$

2. $y = \sin^{-1}(\sqrt{\cosh(\log x)})$

$\dfrac{dy}{dx} = \dfrac{1}{\sqrt{1-\cosh^2(\log x)}} \cdot \sinh(\log x) \cdot \dfrac{1}{x} = \dfrac{\sinh(\log x)}{x\sqrt{1-\cosh^2(\log x)}}$

3. $y = \log(\cosh(\sqrt{\sin^{-1}x}))$

$$\frac{dy}{dx} = \frac{1}{\cosh(\sqrt{\sin^{-1}x})} \cdot \sinh(\sqrt{\sin^{-1}x}) \cdot \frac{1}{2\sqrt{\sin^{-1}x}} \cdot \frac{1}{\sqrt{1-x^2}}$$

$$= \frac{\sinh(\sqrt{\sin^{-1}x})}{2\sqrt{(1-x^2)\cdot \sin^{-1}x} \cdot \cosh(\sqrt{\sin^{-1}x})}$$

4. $y = \cos(\log x) \cdot \log(\cos x) \cdot e^{\sqrt{\cos(\log x)}}$

$$\frac{dy}{dx} = \cos(\log x) \cdot \log(\cos x) \cdot \left[e^{\sqrt{\cos(\log x)}} \cdot \frac{-\sin(\log x)}{2x\sqrt{\cos(\log x)}} \right]$$

$$+ \cos(\log x) \cdot e^{\sqrt{\cos(\log x)}} \cdot (-\tan x) + \log(\cos x) \cdot e^{\sqrt{\cos(\log x)}} \cdot \left(\frac{-\sin(\log x)}{x} \right)$$

5. $x = a \log[\tan(\frac{\pi}{4} + \frac{\theta}{2})]$, $y = a \sec \theta$

$$\frac{dx}{d\theta} = \frac{a \sec^2(\frac{\pi}{4} + \frac{\theta}{2})}{\tan(\frac{\pi}{4} + \frac{\theta}{2})} \cdot \frac{1}{2} = \frac{a}{2 \sin(\frac{\pi}{4} + \frac{\theta}{2}) \cdot \cos(\frac{\pi}{4} + \frac{\theta}{2})}$$

$$= \frac{a}{\sin(\frac{\pi}{2} + \theta)} = \frac{a}{-\cos\theta} = -a \sec\theta$$

V. Find the derivative of the following w.r.to x

1. $x^5 \tanh x - \sqrt{x} \cdot \cot x$
2. $5 e^{\tan x} + \cos^{-1} 2x \cdot \sin(\sin x)$

3. $\frac{x^2}{a^2} + \frac{y^2}{b^2} = 1$
4. $\frac{\sin x}{1+\tan x}$
5. $\frac{x^3 e^{5x} \sec x}{x^2 - 1}$

Answers to the Self evaluation test

I 1. c   2. c   3. a   4. c   5. a

   6. c   7. b   8. c   9. a   10. c

II 1. c   2. a, c   3. b   4. d   5. d

III 1. $\sec^2 x$, $(1 + \tan x)^2$ 2. $\log_a e$, $\frac{2\sec^2 2x \cdot \log_a e}{\tan 2x}$

   3. $\tan\theta$, $\tan^{-1}x$, $\frac{\tan\theta}{\sqrt{1+\tan^2\theta}}$ 4. $a \sinh\theta$, $b \cosh\theta$, $b \tanh\theta$

5. $\log \sin \frac{x}{2} + \log \sin \frac{x^2}{2^2} + \log \sin \frac{x^3}{2^3} + \log \sin \frac{x^4}{2^4}$;

$\frac{1}{y} \cdot \frac{dy}{dx}$; $2x \cdot \cos \frac{x^2}{2^2}$; $2^3 \sin \frac{x^3}{2^3}$

IV Corrected Solution

1. CS : $\frac{dy}{dx} = \sin(e^{\sqrt{\tan x}}) \cdot \sec^2(\sqrt{e^{\sin x}}) \cdot \frac{1}{2\sqrt{e^{\sin x}}} \cdot e^{\sin x} \cdot \cos x$

$+ \tan(\sqrt{e^{\sin x}}) \cdot \cos(e^{\sqrt{\tan x}}) \cdot e^{\sqrt{\tan x}} \cdot \frac{1}{2\sqrt{\tan x}} \cdot \sec^2 x$

2. CS : $\frac{dy}{dx} = \frac{1}{\sqrt{1-(\sqrt{\cosh(\log x)})^2}} \cdot \frac{1}{2\sqrt{\cosh(\log x)}} \cdot \frac{\sinh(\log x)}{x}$

$= \frac{\sinh(\log x)}{2x\sqrt{\cosh(\log x)[1-\cosh(\log x)]}}$

3, 4 - No error

5. CS : $x = a \log[\tan(\frac{\pi}{4} + \frac{\theta}{2})]$, $y = a \sec \theta$

$\frac{dx}{d\theta} = \frac{a \sec^2(\frac{\pi}{4} + \frac{\theta}{2})}{\tan(\frac{\pi}{4} + \frac{\theta}{2})} \cdot \frac{1}{2} = \frac{a}{2 \sin(\frac{\pi}{4} + \frac{\theta}{2}) \cdot \cos(\frac{\pi}{4} + \frac{\theta}{2})}$

$= \frac{a}{\sin(\frac{\pi}{2} + \theta)} = \frac{a}{\cos \theta} = a \sec \theta$

$\frac{dy}{d\theta} = a \sec \theta \cdot \tan \theta \implies \frac{dy}{dx} = \tan \theta$

V Solutions

1. Let $y = x^5 \tanh x - \sqrt{x} \cdot \cot x$, then

$\frac{dy}{dx} = x^5 \cdot \frac{d(\tanh x)}{dx} + \tanh x \cdot \frac{d(x^5)}{dx} - \sqrt{x} \cdot \frac{d(\cot x)}{dx} - \cot x \cdot \frac{d(\sqrt{x})}{dx}$

$= x^5 \cdot \operatorname{sech}^2 x + 5 \tanh x \cdot x^4 + \sqrt{x} \cdot \operatorname{cosec}^2 x - \frac{\cot x}{2\sqrt{x}}$

CALCULUS I – Differential Calculus Formulae Practice Workbook Vol II

2. Let $y = 5\, e^{\tan x} + \cos^{-1} 2x \cdot \sin(\sin x)$, then

$$\frac{dy}{dx} = 5\, e^{\tan x} \cdot \sec^2 x + \cos^{-1} 2x \cdot \cos(\sin x) \cdot \cos x - \frac{2\sin(\sin x)}{\sqrt{1-4x^2}}$$

3. $\dfrac{x^2}{a^2} + \dfrac{y^2}{b^2} = 1$ can be written as $\dfrac{1}{a^2} \cdot x^2 + \dfrac{1}{b^2} \cdot y^2 = 1$

Differentiating w. r. to x, $\dfrac{1}{a^2} \cdot 2x + \dfrac{1}{b^2} \cdot 2y \cdot \dfrac{dy}{dx} = 0$

$$\Rightarrow \frac{dy}{dx} = -\frac{b^2 x}{a^2 y}$$

4. Let $y = \dfrac{\sin x}{1+\tan x}$

$$\frac{dy}{dx} = \frac{(1+\tan x)\cdot \cos x - \sin x \cdot \sec^2 x}{(1+\tan x)^2}$$

$$= \frac{\cos x + \sin x - \sin x \cdot (1+\tan^2 x)}{(1+\tan x)^2}$$

$$= \frac{\cos x - \sin x \cdot \tan^2 x}{(1+\tan x)^2}$$

5. Let $y = \dfrac{x^3\, e^{5x} \sec x}{x^2 - 1}$

Taking log both sides, $\log y = \log x^3 + 5x \log e + \log \sec x - \log (x^2 - 1)$

Differentiating w.r.to x, $\dfrac{1}{y}\dfrac{dy}{dx} = \dfrac{3}{x} + 5 + \tan x - \dfrac{2x}{x^2 - 1}$

$$\frac{dy}{dx} = \frac{x^3\, e^{5x} \sec x}{x^2 + 1}\left(\frac{3}{x} + 5 + \tan x - \frac{2x}{x^2 - 1}\right)$$

CALCULUS I – Differential Calculus Formulae Practice Workbook Vol II

To the students ……..

❶ Student : After successfully completing the practice given in volume I and II, now my fear about the symbols and notations is gone. Moreover, I have the confidence of solving any problem in Differential Calculus. There is no doubt that these methods will help me to get more marks in the examinations. But still something is irking in my mind. I have the following questions.

1. Whether the symbol $\frac{dy}{dx}$ is a 'single thing' or it is $dy \div dx$ (a ratio)?

2. $\Delta x \to 0$ means $\Delta x = 0$ or not?

3. We have proved 'If $y = \sin x$ then $\frac{dy}{dx} = \cos x$' and other similar results in volume 2 using the 'first principle technique'. But these results seem to have no life at all!

Explanation :

1. The answer is in your question itself. You have stated that $\frac{dy}{dx}$ is only a **symbol**. As such **it is not a ratio or fraction**. Hence $dy \div dx$ is not true.

But $\frac{\Delta y}{\Delta x}$ **is a ratio**. Try to recollect the real life example 'the growth of a plant' given in volume I. Δh and Δt are two small quantities and hence $\frac{\Delta h}{\Delta t}$ is a ratio. ie. Δh can be divided by Δt.

We define 'the differential coefficient of y w.r.to x' only when Δx and Δy are infinitesimally small. [ie. when $\Delta x \to 0$ and $\Delta y \to 0$]. This stage is known as the 'limiting stage'.

$\frac{dy}{dx} = \lim_{\Delta x \to 0} \frac{\Delta y}{\Delta x}$. ie. The limiting stage of $\frac{\Delta y}{\Delta x}$ is denoted by the symbol $\frac{dy}{dx}$.

Is it clear now?

2. In the real life example mentioned above, $\Delta h \to 0$ means 'growth is there; the rate of the growth of the plant is infinitesimally small, but not zero'. Hence $\Delta h \neq 0$.

3. These results have life! There is no doubt. Read the following

3.1 Geometrical application

Each function and its derivative have geometry behind them. The limiting process plays an important role in geometry.

The tangent at a point on a curve y = f(x) is the limiting process of a number of secants to the curve passing through that point. You can find this concept very well explained in your text books.

In **analytical geometry**, every line is identified by its slope. Assume that a line makes angle θ with the positive direction of the x-axis, its slope is given by $\tan \theta$.

In **differential calculus**, the limiting process of the secants passing through a given point on the cure y = f(x) give the slope of the tangent at that point.

Hence, slope of a tangent = $\tan \theta = \lim_{\Delta x \to 0} \frac{\Delta y}{\Delta x} = \frac{dy}{dx}$

The above is a geometrical application of the derivative of a given function.

Also derivative of a given function has physical applications too!

3.2 Physical application

Every one of us are growing. This growth is seen in our height, weight ….etc. Trees and animals are growing. Science tells us that our universe too is expanding at a constant rate. Isn't it?

Now we are very eager to know the instantaneous rate at which our height or weight is increasing. Then there should be a systematic method to find out this rate of growth. Am I correct?

Well! This has been explained in the real life example 'growth of a plant' given in volume I.

$\frac{dh}{dt} = \lim_{\Delta t \to 0} \frac{\Delta h}{\Delta t}$ gives the instantaneous rate at which our height is increasing;

$\frac{dw}{dt} = \lim_{\Delta t \to 0} \frac{\Delta w}{\Delta t}$ gives the instantaneous rate at which our weight is increasing;

As distance (s) and time (t) are involved, the function of a moving car can be represented by the equation s = f(t); See the beauty the equation s = f(t) has life!

$\frac{\Delta s}{\Delta t}$ gives the average velocity. The instantaneous velocity of the moving car at any time t is given by $\frac{ds}{dt} = \lim_{\Delta t \to 0} \frac{\Delta s}{\Delta t}$. Similarly the instantaneous acceleration of the moving car at any time can be calculated.

Dear students, these concepts are well explained in your text book. At this stage, read the theory given in your text book carefully; refer at least two other differential calculus books. You can easily understand the life behind theory. You can do it!

❷ **Student :** Look at my teacher who is handling Differential Calculus to us! How confident and accurate he is, while solving problems. I wish to be like him. How to become a master in solving Differential Calculus problems?

The answer to this question will be available in **Prof MSDOSS Maths book series V - 'Differential Calculus formulae practice workbook Volume III'** (available soon)

Wish you all the best!

Available soon

PROF. MSDOSS MATHS BOOK SERIES V

MATHEMATICS

CALCULUS I

DIFFERENTIAL CALCULUS

FORMULAE PRACTICE WORKBOOK - Vol III

FOR THOSE STUDENTS WHO WANT TO BECOME A MASTER IN DIFFERENTIAL CALCULUS FORMULAE & VARIOUS METHODS

www.ingramcontent.com/pod-product-compliance
Lightning Source LLC
Chambersburg PA
CBHW081250180526
45170CB00007B/2361